S.T.E.C. II

Who Would You
Talk To?

by

James W. Steines

Copyright 2016

Written by
J. W. Steines

Dedication

This book was written for everybody in this human race who has lost a loved one, or a dear friend, who they would like to speak with just one more time. The dream of having a morning cup of coffee with a person who has passed on could become a reality.

Introduction

"It is very beautiful over there."

Death is inevitable. It has happened to billions of people in the past, and will happen to billions more in the future. It may not be possible to speak to anybody living hundreds of years into the future, (yet), but it is now possible to speak, face to face, to a person who has lived in the past.

Using a combination of science and religion, a way has now been discovered to allow a living person to carry on a personal conversation with a person who had lived and died before our time. Please understand there will be references to the Holy Bible in this book that I have used in parts of my research.

To learn the truth about history, one must be able to speak with a person who had lived during that period of time. Obviously, it was not possible to speak with anybody who had lived beyond a hundred years into the past. Until now ...

The STEC is short for the 'Spiritual Tele-communication Center', which is a machine which allows the user to contact and carry on a conversation with a person who has passed away.

There are many parts to the STEC I need to create or purchase, thus the reason for having this book written. Even though I have been able to create many of the mathematical calculations for the STEC, there are many more left to complete.

Along with the different computer programs I need, there are also many physical pieces of the STEC which need to be purchased.

I thank you for purchasing this book, allowing me to continue the building and construction of the STEC.

Chapter One
Speak to Spirits

My name is James W. Steines. I am the creator and inventor of a new machine I call the 'Spirit Tele-communication Center' or 'STEC' for short. This machine would allow any person to call for and speak with the spirit of a person who lived and died in the past.

It has been said that science and religion don't mix. At this time, I would have to disagree with that statement. In order for the STEC to work, one must believe that there are spirits existing in another dimension. The project for me was to find a way to enter that dimension and to have the spirits enter into ours.

I began this project in 1995, never realizing I would be making this a lifetime dream. In the past few years, I also noticed the numbers in my bank account were beginning to dwindle. I started contemplating if I wanted to continue working on the STEC, or put it aside until I had enough finances to continue working on it.

In 2006, I was fortunate to attend a presentation by a brilliant physicist promoting his own book to raise funds to continue his own research for his invention. After enjoying his presentation, I was not only proud to purchase one of his books (which he signed for me), but also toyed with the idea of doing the same thing to raise funds for my own invention.

I have to admit, growing up, and even in my teenage years, I never was a big reader. It was not until recently that I have begun to understand and respect anyone who had the ability and the knowledge to write their books for the public to enjoy.

After reading quite a few books, trying to get an idea of how to approach my own, I came to the conclusion there were hundreds of different writing styles that authors have chosen to use. There were, however, two series of books I have come to enjoy. I contacted these two authors in hopes of even getting them to be kind enough to answer my e-mails. I was honored when one of them did.

Not being a writer myself, I tried to seek the advice of one of my favorite independent authors, hoping he would be kind enough to give me a few tips of how to write a story that many will enjoy. He was nice enough to take timeout of his busy schedule to help me prepare this book. I am forever grateful for his help.

I explained a little of my project to Louis, hoping he would understand what I was attempting. To my surprise, he did. Allow me to introduce you to him.

In August of 2012, Louis C. Douglas introduced the first book in his series called *'Speak 2 Spirits - Dragon Tree'*. I found this book quite by accident when I was searching for something to read. Initially, I was hoping to find a book on scientific ways to talk with the spirits. Instead, I found this very fun series and found myself becoming lost in the fictional world he created. Although the series was originally written for a younger generation to enjoy, I have found myself reading the books over again, patiently waiting for the next one in the series to come out.

I contacted Louis, told him what I was thinking of doing and asked if he would have time to talk with me. I was honored when he said he would be more than happy to help.

I became an instant fan of his, immediately sending off for the second book of the series. A few months later, I heard he was having a book-signing a couple of hours away from me. I knew it would be a chance to meet him, and hopefully get his autograph on the first two books I would have.

The day finally came and I had my copies of his books ready for him to sign. I left home a little bit early, hoping to be able to see him before the crowds began to gather for his signing. I was fortunate to have a few minutes to speak with him about my idea before his fans began to come in to see him. In that short time, a friendship was struck and we have been in touch ever since.

Mr. Douglas is an independent author, which means that he does everything on his own to make the books available to the public. Not only does he take the time to write the books, but also self-publish them and promote them to the world. This is the type of people I admire with their persistence and dedication to the thing they love to do. I thank him for sharing his knowledge and taking the time to speak with me. I wish him the best of luck in his literary ventures.

There are two main points of writing which many of these authors agree on. First, write what you love and love what you write. Not everybody is going to enjoy your writing, but there are going to be many that will.

Secondly, keep it simple. Nobody likes to be reading a book and having to stop to get a dictionary out to find the meaning of a word they don't know. Even though you are writing for adults, there will always be young people who might take an interest in your work. You _must_ keep a reader interested in what you are writing about.

The following story is the story of how I came up with the idea of creating a machine which would allow me to sit down across a table and have a conversation with a person who had passed away. Although I enjoy a good scientific book, I also enjoy a good book with a fun storyline.

I spoke with Louis, asking him for his advice and was grateful when he agreed to help me with this new undertaking. We decided to make this a fantasy book with my research and findings intertwined within the story.

Obviously, the book will be partially fiction, including the conversations I will be writing about with a few famous people. I will be including excerpts from some of the interviews they gave to publications or television.

The names of my friends I will be using are going to be changed to protect their identities. To the families of these friends, I want to let you know that I loved them both dearly.

My research and findings will be real, as well as the mathematical equations I use. But of course, I will not be putting in all of the information of how the machine functions as to keep this invention my own. In this book, I will be including a few of the theories and applications I will be using in the research for the STEC.

I hope you will enjoy the rest of the book.

Chapter 2
Grandpa's Lesson

The snow began to fall during a trip to visit my grandparents in Montana. In my young eyes, Grandpa was the most talented craftsman in the world. At the time, I thought he could build a house out of a single toothpick.

During this snow day with him, he took me to his woodshop for the very first time. I must have been about seven at the time and remember thinking that this had to be the greatest place in the world. I remember smelling the aromas of the different types of woods in the air as Grandpa and I walked through the doors.

Grandpa must have seen the excitement in my eyes and let me look around the entire room, telling me the name of the different tools whenever I asked him what they were. I enjoyed listening to Grandpa talk since he was born and raised in the deep south and still had his strong southern way of speaking.

"What is this?" I asked, pointing at one of his many tools on his workbench.

"That there's a chisel," he said, watching me look it over carefully. "That be one that I use to make fancy designs in the wood."

I looked around and realized that he had to have at least fifty different shapes and sizes of chisels. He quickly explained that he needed the different sizes to make the different carved designs into the wood. I think he told me that so I wouldn't end up asking him what each and every chisel did.

After a half hour of answering my questions, he finally said those magic words I was hoping to hear.

"Do ya wanna build something with me?" he asked, watching the excitement in my eyes begin to show.

"Yeah," I said, following him over to his workbench. I watched as he reached for a leather apron that was hanging on a nail on the wall and a small pair of safety glasses.

"Whenever you work with wood," he said, putting the apron on me and tying it in the back, "You have to make sure you wear protection. I don't wanna have to take you to the hospital 'cause you hurt yourself, okay?"

"Okay."

"Now," Grandpa said, reaching over for a hammer and a nail, "A good builder has to know how to hammer a nail into a board. Can you do that?"

I shook my head no as I put on my safety glasses. After getting me a stepstool to stand on to watch him better, Grandpa grabbed a small board from the top of the table and put it in front of us, so I could watch him pound a nail into it.

I watched as he firmly held onto the nail with his left hand and tap it with a hammer he was using in his right. After a few taps, he let go of the nail, letting me see that it was deep enough into the wood to stand up on its own.

"That's the easy part," he told me, showing a half smile. "The hard part is making sure the nail goes in straight and not crooked."

With one swift swing of the hammer, he hit the nail right on the top, driving it down flush with the board.

"Now, you try it," he said, giving me the hammer and another nail.

I leaned over the board and put the nail about an inch away from his. Hitting it softly, it went down far enough to stay in the board without leaning too badly.

"Go for it," Grandpa said. "Smack it good now."

Grandpa moved back a couple of feet as I raised the hammer high into the air, bringing it down towards the head of the nail. We both began to laugh when the hammer struck the nail, sending it across the room in front of us, hitting a pile of two by fours leaning up against the wall.

"Okay," Grandpa said as we both began to stop our laughing, "There are two things you have to work on. First, maybe you shouldn't try to pound the nail in with just one hit. Just keep hitting the head of the nail until it goes in. Secondly, you ain't holding your mouth right, boy."

I looked over at him and chuckled when he showed me how he wanted me to hold my mouth the right way. Softly tapping in another nail, he squinted his eyes tightly and stuck out his tongue. Raising the hammer, he once again hit the nail square on the head, driving it deep into the board.

After about a dozen more nails, I was finally able to pound a nail straight into the board. Grandpa was proud of me, even though I bent eleven of the twelve nails he gave me to practice with.

I practiced hammering more nails into the board while we decided what we were going to build together. I asked Grandpa what he wanted to build and we decided to try to build a birdhouse together.

"There are three things you have to remember before you build something," he said to me, looking at me to make sure I would understand what he was going to tell me. "Conceive it, believe it, and achieve it."

He must have seen the confused look on my face as he began to smile.

"If you can think of an idea and believe you can build it, then you can succeed in making it a reality. But, you have to really want it and never give up on achieving your goals."

"I understand," I told him, knowing what he was trying to tell me.

I spent the next few days with Grandpa building the birdhouse. I was nervous using a few of the machines, including the electric saws we used to cut the wood, but was able to finish our project without losing any fingers or hurting either me or Grandpa.

In 1979, Grandpa passed away. I went to his home after the funeral and walked out into his workshop area to look around. Almost everything was covered with sawdust, except for one crooked birdhouse sitting on one of the many shelves. It was about the only clean thing in the shop. Now, it is sitting on a shelf in my home.

It has been over thirty-five years since Grandpa died, but I can still smile whenever I think of the great time we had building that birdhouse.

Chapter 3
Goodbye My Friends

Throughout my life so far, I have been able to meet and become best friends with two specific people. I met Sam during my freshman year in high school, instantly becoming friends. Through the years, Sam and I became as close as two people could get.

The second was Debbie. Debbie moved to our town while Sam and I were in our junior year at school. Although Debbie was a year behind us, we soon became a tightly knit group. We built this friendship stronger throughout the following years, keeping in close touch with each other, no matter where we ended up.

Sam accepted a job offer in Texas, moving there a couple of years after graduation, while Debbie moved out to the east coast where she fell in love and married a very nice man. I decided to move away and found a beautiful quiet town in Missouri to settle down in. Together, the three of us could have the best of times just sitting in the middle of a restaurant talking. If any one of us were going through a tough time, the others were just a phone call away.

I remember the very day my life took a strange change of direction. It was May 20th, 1992. The sun was just beginning to set when I parked my car in the driveway at home. I had been away for two long weeks for my job and was relieved when I saw my home in front of me. I didn't have to be away from home very often, but when I did, it was difficult.

I fumbled for the front door key as I walked towards the welcoming sight of my home. As I expected, the answering machine was full of messages from the last two weeks, so I decided to wait until I could relax a little from the seven hour trip I just finished. *(You must remember that at this time, we did not have the cell-phones we are use to having now.)*

I walked back out and unloaded the car, bringing in everything I had to take with me out of town. After the third load of stuff, I walked in and sat down in my favorite recliner. That was a mistake, as I found myself waking up about an hour later from a well deserved nap when the telephone began ringing.

"Hello?"

"Hi, Jim? This is Sam." The voice on the other end said.

"How are you, Sam?" I asked, continuing with the phone call.

"Jim," he said as I noticed a tone of sadness in his voice, *"Debbie died a few nights ago."*

"She what?" I asked, not wanting to believe what I was being told. *"I just talked to her last week. What happened?"*

"Her husband went into their bedroom last Sunday to surprise her with breakfast before they went to church and found her still in bed. He tried to wake her up, but she was already gone. They are saying it was a heart attack."

Sam and I continued talking about Debbie for about another hour before hanging up. It was difficult for me not to recall many of the great times the three of us had. It was even more difficult not to remember the great times Debbie and I had together. There was so much more I wanted to talk to her about, but now I wouldn't ever have the chance.

Allow me to move ahead a few years now. It was coming close to the five-year anniversary since Sam and I lost our good friend, Debbie. Times had changed for me and I was beginning to go through a severe mid-life crisis. My work no longer interested me and I began searching for something new to do.

I knew there was something else I could learn, but had a hard time deciding what I wanted to do. One horrible night it came to me when I got a telephone call from another acquaintance of mine, telling me that Sam was just taken to the hospital in his home town. He was involved in a bad car accident and wasn't expected to make it through the night. This was the first time in a long while that I actually bowed my head and prayed for Sam to make it.

I could feel something around me but couldn't see anybody. Suddenly, it felt as if I was being given a gentle hug from this invisible force. I never experienced anything like this before, and unfortunately, I haven't felt it since that time. I would find out later that Sam passed away just minutes before I felt that invisible hug. Somehow, I just knew it had to have been him stopping by to say good-bye to me one last time.

That night as I laid in my bed thinking about Sam and Debbie, I knew there had to be a way that I could speak with them again. I have always been told that when a person passes and makes it to Heaven, that all of their family and friends would be there to welcome them. I didn't want to wait that long before I could see Sam and Debbie, so I tried to think of a way I could see them and talk with them while I was still alive.

I thought of different ways to speak with my two friends, although none of them were feasible to accomplish. One night about a year after Sam passed away, an idea came into my head. Once again, I shook it off as being impossible to build. This invention would not leave my thoughts as the others did, making me think it might just be possible to create. I finally found that one thing I was searching for to satisfy my mid-life crisis.

I was now ready to research and build my machine that would allow me to have a conversation with my two best friends, both who have passed on after being a big part in my life. The big question now was, what should I be researching? Suddenly, I remembered something my grandfather told me long ago.

"Jim," he said, *"There are three things you must remember if you want to be a success in your lifetime. It is very easy to remember. 'Conceive it, Believe it and Achieve it.' If you can think of something and believe that you can do it, then you will be able to achieve that goal."*

I have conceived the idea of a machine that would allow me to speak to the spirits of my two friends, now I have to convince myself to believe that it is possible to build successfully. There are many subjects I would have to search for on the internet, but the first would be actual photographs of spirits.

I always believed there were spirits floating around the earth, but up until this point in my life, I was never able to see one in person, sort of speak. I looked on the internet for pictures of spirits, or ghosts, and was amazed to find thousands of different websites that had photo's of spirits. I leaned back in my chair and tried to figure out where to begin with these sites.

I knew there was computer software that could be used to altar photographs, therefore creating pictures with ghosts in them, so I decided to limit my search for pictures from ten years to a hundred years b.c. (before computers). This reduced the number of photographs drastically.

I was now convinced that the spirits were around and they could be seen. Now the question becomes, can a person actually hear the spirits when they are talking with them?

Chapter 4
Ghosthunter Richard

I had the pleasure of meeting a man by the name of Richard who was a modern day ghost hunter. I wasn't quite sure what he did, so he was kind enough to explain it to me.

He explained that his group would visit homes and business's where the owners thought a ghost would be haunting. They would take many electronic devices with them, such as digital voice recorders, infra-red cameras and many other pieces of specialized equipment. They would spend the entire evening at the place in hopes of capturing any sign of the spirits.

Richard went on to say that most of the time they would not get any proof of spiritual activity, but there were a few times they would. He was kind enough to show me a few photographs of what they thought were spirits, but I could not see their features clearly. This, he told me, is another reason some of the ghost hunters actually bring their dogs along with them. There have been a few instances where the dog would stop and listen to what might have been a spirit in their presence. After listening to the audio replays, a different voice could be heard talking at the same time the dog stopped.

I turned my focus of his investigation to the audible part and asked if his team was able to record any voices that could not be explained. He invited me over to his home that evening to listen to some of the recordings he and his team were able to capture. The first time I heard a voice from the spirits, the hair on my arms and neck stood straight up.

Curious about the recordings, I began asking many questions. He assured me there were no other people around, with the exception of his group, and the video recordings could verify that.

Richard also told me that even though it is possible the recordings could have been from a radio station frequency, the clips I heard were not.

"The voice clips you are listening to," Richard told me, "are what's known as 'direct responses'. Those are answers to questions we were asking at the time."

He allowed me to listen to a few of the direct responses, letting me listen to a team member ask the questions and hearing the replies from the spirits. He admitted that they could not hear the answers at the time, but could hear them plainly when they listened to the recordings.

"The spirits reside on a different level of existence than the living," Richard explained. "There are times when the two levels combine, allowing the living to see the spirits and the other way around."

I became intrigued about the possibility of capturing my own spirit voices and talked with Richard many more times to learn how I could do this on my own. We talked about the tools I would need, letting me know how expensive it was if I wanted to get all of the equipment he and his team used.

After listening to more of the recordings and talking with him, I thanked him for his time and went out to continue my own research without the help of a team like he had. Richard suggested that to begin, I should purchase an electronic voice recorder to use. He smiled when he told me it did not have to be an expensive one, but just good enough to record sounds from a distance.

I went out the next day to an electronic store and bought a fairly inexpensive recorder to begin my research with. After looking around, I found a small out of the way cemetery for my first try to capture some voices. I took along my digital camera to take some pictures, just to see if I could get lucky enough to capture my own pictures of a spirit roaming around the cemetery.

I looked around the area where I live and found a peaceful cemetery only about a half hour from where I am at. *(I will not say where this cemetery is at, so to keep it a peaceful place for those who are buried in it.)* I drove over one sunny day, parked my car and walked silently through the area, trying to see if I could hear any other voices nearby.

With my camera and my recorder turned on, I walked into the cemetery for the first time to attempt to capture voices from the graves. I found there was such a different sense of peace and tranquility as I walked through the cemetery and read the dates and names on the tombstones.

Richard told me that in order to capture spirit voices; I would have to be the one to initiate the conversation. There have been many times I talk to myself out loud, but this time seemed different when I realized I would be trying to talk to invisible entities.

I quickly grew accustomed to talking out loud into the air, hoping one of the spirits in the graveyard would answer me back. After about an hour of talking, I thanked those who might have talked back to me but I couldn't hear them with my normal hearing.

After my time in the graveyard, I went over to my favorite diner close to the cemetery and went in for a quiet lunch before I went home. My favorite hostess welcomed me and took me to a table in the corner. Many of the other patrons said hello to me, knowing who I was because of the number of times I have visited the restaurant.

Sitting in the corner booth and enjoying my lunch, I suddenly came across an idea for my research. I recalled the very first time I came into this restaurant, nobody knew who I was. After a few more times, a few of them began saying hello to me.

I started wondering if I continued visiting the same cemetery if the spirits of the people buried there would become comfortable enough with me to begin talking with me. Shortly after I went back home, I made a note to remind me to continue visiting the same cemetery on a weekly basis.

Although I only recorded about an hour of my time searching for voices, I was surprised when it took me almost three hours to listen to that first recording. Richard told me there would be many sounds I would have to listen to over and over again to see if it was a voice or something different.

After listening to the full hour of a recording, I was surprised to have been able to find three voices that I did not hear at the time. I visited the cemetery for the entire year and collected many voice clips, including a few of them which Richard called 'direct responses'.

For the next year, I visited the same cemetery on a weekly basis. It seemed my thoughts were correct as the number of voices I recorded seemed to grow in numbers. I believe the spirits were becoming relaxed with me, letting me know they were there. It took me a while, but I have been able to recognize a few of the spirits who "talk" to me whenever I enter the area as the same spirits who had spoken to me in previous visits.

I now had two of my questions answered. Number one was the proof of spirits which I found in old photographs, and number two is that a person would be able to record the voices of those who have passed.

Becoming convinced that spirits do exist, I began the next part of the research to figure out how I could actually see the spirit I was speaking with.

Chapter 5
Sights and Lights

While talking with my ghost-hunter friend, Richard, he mentioned that some teams of researchers would sometimes take their dogs with them to find the spirits. This seemed strange to me until he explained the reason why.

While I was searching for authentic photographs, I accidentally came across a website which claimed animals were able to see spirits. This intrigued me and I began researching this area of my project. Could it be proven that animals actually had that ability?

According to Richard, there are many teams of other hunters who take their dogs out in the field with them. It is uncanny the way the dogs would change their attitudes when they felt, or could hear a spirit. Is it because they can feel the spirit or could they actually have seen one?

The first thing I would have to find is the differences between the sight of humans and animals. Thanks to the internet, I could find the details of this subject quickly. *(Just a side note, I used the internet quite a bit for my research.)*

Visible wavelengths from about 300 to 1400
300 – 400 nm = near ultraviolet wavelength
400 – 420 = wavelength of violet light
420 – 440 = wavelength of indigo light
440 – 500 = wavelength of blue light
500 – 520 = wavelength of cyan light
520 – 565 = wavelength of green light
565 – 590 = wavelength of yellow light
590 – 625 = wavelength of orange light
625 – 700 = wavelength of red light
700 – 1400 = wavelength of near-infrared radiation

In terms of frequency, this corresponds to a band in the vicinity of 430 to 790 terahertz

----

The average human being has the color range between about 390 to about 700. There are some who can see below the normal range or over, but those are very few.

Humans have the three kinds of cone pigments, ones that can detect red, green and blue lights. It's no coincidence that we find these colors in the screen of an average television set. The ability of an organism to see in 'color' thus depends upon the color receptors present in the retina. People (or animals) lacking a specific color receptor are unable to 'see' that color. Most often, it appears grayish, or as one of the other colors that can be detected (i.e. purples appearing grayish-blue).

Dogs can't see the full range of the color spectrum, but that doesn't mean they can't see colors at all. Dogs can't tell the difference between orange, yellow and green. All of these colors look yellowish to a dog. The color red appears dark brownish gray or black to dogs.

Dogs can't see the color blue, but violet shades appear blue to them. Blue shades appear gray to dogs. Therefore, dogs are not completely colorblind, they do in fact display partial colorblindness that inhibits them from perceiving shades of green and red. This type of colorblindness also occurs in humans and is known as deuteranopia.

Researchers believe that dogs can distinguish shades of gray indistinguishable to humans. This is because the canine retina contains a larger number of the rod cells that perceive shades of gray.

From what we have learned here, in order to be able to see the spirits of the afterworld, the range of vision would have to be at the lower part of the spectrum. This would be from 300 to 500 nm. On doing a few tests, I have concluded the lighting of the machine would have to be provided by a ultra-indigo lighting system and using a black and white recording process.

I could tell from the results I gathered, I had many more tests I must perform to make any final conclusions about this aspect of my research. After speaking with the caretakers of the cemetery I visit, I received permission to stay in the graveyard after the sun went down to perform more of my tests.

After borrowing a night vision camera from Richard and setting up some dark lights in the cemetery, I began talking into the air, hoping to start up a conversation with any spirit that might have been around. About an hour later, I packed up everything and returned home to view the footage from the camera.

Unfortunately, I wasn't able to capture any movement in the cemetery except for a couple of low flying bats. I was, however, able to capture a few more voices which were direct responses to the questions I was asking.

I studied the film footage I took in the cemetery many times, hoping to see something I had missed before. Each time I watched it, all I could see interrupting the footage were the two bats.

Sitting back in the chair, a little disappointed at my first attempt to see spirits, I decided to watch the footage once more. This time I changed from the regular view to the negative view.

I was surprised to see a little more movement in the footage than I saw in the normal view. It was difficult to make out but I was sure there was a quick movement going from the right side of the screen to the left. Oh, and I also saw the same two bats fly by again, this time as white bats instead of black.

Chapter 6
Thomas Alva Edison

As I continued my research for ideas to build the STEC, I was surprised to find out that there had been another who had planned to make the same type of a machine. His name was Thomas Alva Edison.

"It is very beautiful over there."

These were the last words of one of America's greatest inventors. Thomas Alva Edison. Edison died on October 18, 1931. Many years prior to his death, he began work on a device to speak to the spirits of those which had passed on before him. This machine was referred to as the "T.E.C.", the 'Thomas Edison Communicator', or otherwise known as the "Valve".

This device would allow him to contact and carry on conversations with the spirits on the other side. There are two common beliefs which had caused Mr. Edison to create this device. One of those beliefs was that he wanted to contact a friend of his who had died a few years earlier, and the other being that he wanted to, once again, speak with his mother who had died in 1871. Although no drawings or schematics for this invention have been discovered, it is believed that he was actually in the process of making a machine he called the 'Valve'.

In a 1920's essay, Edison wrote: *"Now what I propose to do is furnish psychic investigators with an apparatus which will give a scientific aspect to their work. This apparatus, let me explain, is in the nature of a valve, so to speak. That is to say, the slightest conceivable effort is made to exert many times the initial power for indictive purposes. It is similar to a modern power house, where man, with his relatively puny one-eighth horse-power, in that the slightest effort it intercepts will be magnified many times so as to give us whatever form of record we desire for the purpose of the investigation. Beyond that, I don't care to say anything further regarding its nature. I have been working out the details for some time, indeed, a collaborator in this work died only the other day. In that he knew exactly what I am after in this work, I believe he ought to be the first to use it if he is able to."*

In another interview in the October, 1920 issue of "The American Magazine", Edison says, *"I have been at work for some time building an apparatus to see if it is possible for personalities which have left this earth to communicate with us."*

In another interview with Edison, published in the same month and year, in the Scientific American magazine, he said, *"I have been thinking for some time, of a machine or apparatus which could be operated by personalities which have passed on to another existence or sphere."*

Each of these interviews were in October of 1920, yet in one he says he has been 'thinking' of the machine, while in the other, he states that he has been 'building' it. If it is true that he was already building the machine, then a person could assume there was already a prototype somewhere that is yet to be found.

Throughout the interviews he gave, Edison didn't subscribe to the conventional notions of life after death. He surmised that life was indestructible and that *"our bodies are composed of myriads and myriads of infinitesimal entities, each in itself a unit of life. There are many indications that we human beings act as a community or ensemble rather than as units. That is why I believe that each of us comprises millions upon millions of entities, and that our body and our mind represent the vote or voice, whichever you wish to call it, of our entities. The entities live forever. Death is simply the departure of the entities from our body."*

"I do hope that our personality survives," he said. *"If it does, then my apparatus ought to be of some use. That is why I am now at work on the most sensitive apparatus I have ever undertaken to build, and I await the results with the keenest interest."*

Realizing this man's fantastic track record, and his incredible mind, it would be difficult to imagine how different the world would be if he had actually succeeded in building his apparatus.

On February 11, 1938, a memorial was dedicated to the late great Thomas Alva Edison in the Menlo Park area of Edison, New Jersey. The tower is located on the area where Edison's laboratory was built.

Over four hundred of his most important inventions were created in this one lab. The towers pinnacle is meant to represent an incandescent light bulb and originally included an audio system that could be heard from a distance of two miles away.

The tower is 131 feet tall and is topped by a fourteen-foot eight-inch tall bulb made of Pyrex segments from the Corning Corporation.

In 2015, the many incandescent light bulbs that once filled the top of the tower were replaced with LED lighting. In 2012, a major restoration began on the museum and was opened back to the public on October 22, 2015.

At this time, I am going to add my thoughts about the TEC invention Mr. Edison was working on. Knowing the brilliance of this man, is it possible that he *did* actually create this invention? Could it have been a success, but he didn't think that the world would be ready for it?

Mr. Edison was a brilliant man with a great mind. I am having a difficult time believing that if he was working on this invention for well over ten years that he was not able to complete it.

Chapter 7
Do's and Do not's

1 - The wrong way to use the STEC II machine.

When the machine is completed, I will be able to speak with a spirit. I will use a picture of the person I want to speak with, dated 1945. The spirit arrives and I ask a few questions to make sure it is the right one. After becoming convinced, the spirit and I have a great conversation before saying good bye.

A couple of weeks later, I use the same picture taken of the person and the spirit arrives. Although we talked before, the spirit would not recognize that we spoke already and we would have to have the same conversation we had before.

1 - The right way to use the STEC II machine.

To talk with a specific person, I must use the pictures, if that is my choice of contact, and use the earliest picture of that person. The next time I try to contact the person, I have to use another picture taken after the one I used for the previous contact.

2 - The wrong way to use the STEC II machine.

When you are able to have the honor of speaking with a spirit, you must treat them with kindness and respect. Do not insult them, as they have the choice whether or not to help you. Do not ask them for the winning lottery numbers, because chances are they wouldn't know anyway.

2 - The right way to use the STEC II machine.

After connecting with the spirit of the one you want to speak with, treat them with the same respect you would treat them as if they were sitting across from you in their human body. Keep your questions professional and direct. Do not attempt to trick them into telling you things they aren't allowed to tell you, such as what your future might hold, because once again, they probably wouldn't know anything about it.

3 – The wrong way to use the STEC II machine

Never begin arguing with the spirit. As I said before, the spirit does not have to stay to talk with you. They can leave at any time if they feel as if they are being used by you.

3 – The right way to use the STEC II machine.

You don't have to accept everything that you hear from the spirit you are speaking with, but do not try to prove them wrong. You are recording the entire conversation and will be able to watch it over and over again. There will be times when the spirit will not, or cannot, tell you what you want to hear. Sometimes they will tell you what you want to hear, without actually telling you straight out.
There may be times when the spirit will give you the answers to your questions, but you will have to work to understand what they are telling you.

4 – The wrong way to use the STEC II machine.

Do not ask the spirit to do things for you. They are here to help you and assist you in what you are doing. Although it might be tempting to ask a spirit to go and spy on someone for you, or go and scare somebody, that just wouldn't be right on so many different levels.

4 – The right way to use the STEC II machine.

You have contacted a spirit to have a conversation with. Don't try, in any way, to have them do something they might get in trouble for doing. If you wouldn't do it, don't ask the spirit to. They probably have a completely different set of rules to live by, so don't make them get in trouble for doing anything wrong.

These are only four of the many different rules to follow. To put it in basic terms, don't treat the spirits as anything but spiritual help and to learn the history of the times they lived in. If you are contacting the spirit of loved ones, I am sure you will have much more to talk about instead of history.

As for me, I can't wait until I can talk with Debbie and Sam again, and many of my relatives, of course. But, I am also interested in the history of the world. It would be an honor to talk with some of the people who actually lived in the past.

In the following chapters, I will be telling you about a few spirits I would talk to, once the STEC II is working at full strength. As I said in the beginning of the book, parts of the book will be fiction. These conversations I will have in the next few chapters will be fiction.

One of my main concerns is, will I be able to continue talking with the spirits I contact on a regular basis, or will I only be able to speak with them one time? It would be great to be able to speak with them more than once, but I have to figure this out before I try to contact Sam or Debbie.

If it is only a one time connection, I would want to make sure I could set aside a full day to speak with my friends without any type of interruptions. So, for the first try, I will contact a spirit I would only want to speak with a couple of times, making sure to figure out if my thoughts would be correct or not about being able to speak to a spirit only one time.

The question now was how do I contact the spirit I want to speak with?

Throughout the time man has been walking on this earth, there has been, literally, millions of people who have lived and died. With all of the spirits of these people roaming the other side, I would need to come up with a way to contact the one spirit I want to speak with.

As I was watching one of those who-dunnit police shows, the answer showed itself to me. On this show, the detectives were trying to find a killer, or a bank robber, or somebody like that. What they showed next was one of those moments where I just leaned back, wondering how I could miss the obvious.

There were two points I learned from the show, and those were about DNA and fingerprints. I began thinking that every spirit would have to be different, therefore creating a unique 'fingerprint'. The second thought I had was if a spirit was in the human form I was seeking to find, would it be possible the DNA of the body would still have any type of a connection with the spirit?

According to my ghost-hunter friend, Richard, there are spirits who remain close to objects they once had, or owned. If I had a ring from my grandmother, could that ring be a connection to her spirit? This is something I will have to look into.

For the time being, I am going to concentrate my research on the 'fingerprint' aspect of the STEC.

Chapter 8
Time and Fingerprints

The definition of time, according to the Merriam-Webster dictionary, is: *a nonspatial continuum that is measured in terms of events which succeed one another from past through present to future.*

There has been a question I have been having a difficult time finding an answer to. Does time have a vibration, or is time itself a vibration? I have researched this question for many years, but am still unable to find an answer.

The closest answer I could find were in the calculations of a man named Max Planck. Planck was a German theoretical physicist who originated the quantum theory, which won him the Nobel Prize in Physics in 1918.

All scientific experiments and human experiences happen over billions of billions of billions of Planck times, making any events happenings at the Planck scale hard to detect.

As of May 2010, the smallest time interval uncertainty in direct measurements is on the order of 12 attoseconds (1.2×10^{-17} seconds), about 2.2×10^{26} Planck times.

I am not planning on using this small of a time span, rather I will be basing some of my own time calculation measurements.

I will be calling these measurements as 'Living Points', or LP for short. The calculations I am using to acquire these LP's will be kept out of this book until the proper time to let the public know through a scientific paper.

The LP will represent the vibrations of time from a person's birth until his death. Of course, it would be helpful knowing the exact second a person was born and died, but the approximation will suffice.

Once the Birth LP is found, it is important to remember that this young of a spirit might not be strong enough to be able to carry on a good conversation. The longer the spirit lived on the earth, the better the chance of a more knowledgeable conversation could be held.

For this example, I will be using the birth and death dates for Thomas Edison to show the LP's I would use to contact his spirit.

Birth – February 11, 1847
BLP – 1946.3061.7740.8585.2171.7043(8)
Death – October 18, 1931
DLP – 1990.5109.6435.5262.0438.5742(8)

BLP-DLP – 44.2954.8108.8210.5212.6548(8)

In order to begin searching for the spirit of a more knowledgeable Thomas Edison, I would have to locate the LP vibration of time after he was a young man from the age of twenty years or older.

Although I would be able to speak to the spirit of Thomas at anytime after it was created, I'm not sure if the knowledge of his entire life would have been learned yet. This would be a question I will have to wait for the answer to after the STEC is working properly and I can do some extra research on.

Since this would be the first time to talk with Thomas, I have decided on a specific age I would want to concentrate on. I have decided to use the LP of his age of around seventy years old. This would allow him to have successfully created many of his inventions and be able to understand what I was attempting to do.

The contact LP I will be focusing on would be:

1977.7904.7780.3177.8285.7787(8)

I then began thinking about adding the possibility of using the M-theory, which are parallel universes. M-theory is the "Membrane Theory" which is a multi-universe theory. If the M-theory is correct, then it is possible that the spirits all live within this universe, yet would be close enough to our world to intercept with it and the spirits would be able to move around with us and speak with us.

What would happen if the spirit universe was able to merge with us, so we could all see them and talk with them whenever we wanted to? We know spirits can be seen and heard, but is it because of certain conditions must be present, or can we break through to the other spirit universe and merge the two ourselves?

I then began to take a look at the eleventh dimension and try to work it into the possibility of opening the rift between the parallel dimensions using the M-Theory.

Theory – If a person creates parallel universes from every decision they make, the average person will create an average of seventy-five million p.u.'s in their lifetime. From the time of creation, billions of humans have lived and died. If we estimate the number of humans who have existed to be around fifty-billion, that would mean the number of parallel universes created would be approximately 3,750,000,000,000,000,000,000.

That large number could also be stated as 3.75 Septillion, which would be 3.75×10^{21}.

Superstring theory – *According to the theory, the fundamental constituents of reality are strings of the Planck length (about 10^{-33} cm) which vibrate at resonant frequencies. Every string, in theory, has a unique resonance, or harmonic. Different harmonics determine different fundamental particles. The tension in a string is on the order of the Planck force (10^{44}*

newtons). The graviton (the proposed messenger particle of the gravitational force), for example, is predicted by the theory to be a string with wave amplitude zero.

Stay with me here, this might be a bit difficult to understand.

As time continues, so does humanity. Out of the billions of people who were born on this earth, the odds of two people being born at exactly the same time are astronomical. Let's say that two people are born at exactly 12:43:13. Even though they were both born at the same second, one of them had to have been born, perhaps, one-billionth of a second before the other. Therefore, the first born would have a different harmonic string than the latter.

As the definition of the string theory states – 'every string, in theory, has a unique resonance, or harmonic'.

Ecclesiastes 12:6-7 states: *"Or ever the silver cord be loosed, or the golden bowl be broken, or the pitcher be broken at the fountain, or the wheel broken at the cistern. Then shall the dust return to the earth as it was: and the spirit shall return unto God who gave it."*

Is the silver cord and the string in the string theory the same? I'm not really sure, but what I do know is that it could be possible.

According to many of the leading theoretical physicists, the strings of the string theory all lead to what is being called the 'M-theory'. The M-theory is described basically as where all of the strings in the string theory end.

Reading the definition of the string and the "M" theories, and comparing it to the Ecclesiastes passage, I can see a similarity between the two. All strings of the theory lead to the membrane of the M-theory. All the silver cords lead back to God. To me, it sounds to be about the same.

If my theory is correct, then it would mean that every human being has a different string connected to God. Each of the strings would have their own vibration, just like each person has their own unique fingerprints; no two would be the same.

Now comes the question, how do I find the one string out of the billions and billions of others, making sure I am able to speak with the spirit of my choosing? There are three possible answers.

1 – DNA
2 – Personal item
3 – Photograph

The size of a string in the string theory = one trillionth of a millimeter

=

1 millimeter = 0.03937008 inches
1 inch = 25.40000 millimeters

Many people who have survived a near-death experience, claims to have seen a bright silver cord which connected their spiritual body to their physical body. Many who have seen it during their experiences say it is about an inch wide and very flexible.

Normal mathematics:

75,000,000 decisions x 0.03937008 inches = 2952756 millimeters = 116250.2362 inches = 9687.5197 feet = 1.8348 miles. That is a mighty large string connecting a spirit to a living person.

Using the estimated size of a string in the string theory at one trillionth of a millimeter and if you take a persons 75,000,000 decisions x 0.00000000000003937008 inches = 0.0000002952756 millimeters = 1.16250236 inches, it would get you very close to the size of the one-inch silver cord the people with their near-death experiences claim to have seen.

Chapter 9
Make-up of a Spirit

In order for me to see a spirit well enough to have a conversation with them, I started thinking of what a spirit would be made of.

What exactly is a spirit of a person? It is said there are three main parts to a human being. There is the body, the soul and the spirit.

The body is a vessel we use so we can survive on this earth. The soul is where all of our information is kept, learning right from wrong, good and bad, storing memories, etc. The spirit is the glue that holds everything together during our lifetime on this planet.

In 1921, Dr. Duncan MacDougall, a physician in Haverhill, Mass., set out to find the weight of the soul when a person died. Although his testing was not regarded as having any scientific merit because the weight loss was not reliable or replicable, his findings were unscientific.

His conclusion was the soul of the dying patient weighed approximately ¾ of an ounce, or 21 grams. My only problem with this result is that Dr. MacDougall refers to this weight loss as the weight of the 'soul' leaving the body.

Not wanting to get into a religious debate or discussion at this time, I will say my belief's here. I was always taught, as I mentioned earlier, that the soul goes back to the heavens and the spirit is permitted to roam the heavens freely (if it is allowed to). So, I am using Dr. MacDougall's theory as the weight of a spirit, not the soul, and will refer to it as such.

Now the thought occurred to me, what would a spirit consist of? If there was truly a weight to a spirit, it had to consist of some type of mass to weigh anything at all.

The body is made up of chemicals which have been painstakingly identified throughout time in the medical field. Although there are sixty different elements in the human body, my only concern for the composition of the spirit are the gaseous elements. The list of chemicals is as follows for a 155 pound human:

Element	% of mass	% of atoms
Oxygen	65	24
Carbon	18	12
Hydrogen	10	12
Nitrogen	3	0.58
Phosphorus	1	0.14

Sulfur	.25	.038
Chlorine	.15	.024
Fluorine	.0037	.0012
Bromine	.00029	.000030
Iodine	.000016	.00000075
Selenium	.000019	.000000045

The total weight of these gases would have to come close to being our average weight of a spirit, which is approximately three-quarters of an ounce.

According to the research of many experts, and even though I am not an expert, myself included, a spirit has a very cold feeling when it touches you or an object. This would leave me to believe that the gaseous state of a spirit must be able to retain a cooler state of consistency.

Even if the spirit was made from these gasses, there would have to be something that would keep the spirit in its proper form. You could call it a 'spiritual shell', if you want. If there were nothing holding these gasses together, they would just float off into space, not being able to be seen.

The question now becomes, what is the glue that would hold the gases together, creating the entity?

Not knowing what the laws of nature are in the spirit realm, it would be difficult to determine what the answer to that question would be. Perhaps, after the machine is invented and working properly, I will be able to find out what some of their laws are from a spirit I would be able to speak with.

Knowing what some of the laws of the afterlife are, would put to ease the minds of many people living on this earth. It could also allow us to begin living our lives better, knowing what lies ahead.

As I said in the introduction, '*to learn the truth about history, one must be able to speak with a person who had lived during that period of time.*' What if we were able to talk with a spirit from many hundreds of years ago and find out that the history we have been taught was wrong?

Let's say, for instance, you had heard that one of your 4th great-grandfathers was walking through the woods one day after a hard day of hunting for food for his family. According to your family stories, he was attacked by a big bear on his way home.

According to the families stories, Grandpa fought off the bear with a stick he found on the ground and killed the bear. During the fight, the bear mauled Grandpa's leg so badly that it had to be taken off. But, Grandpa still killed the bear and with his one good leg, dragged it through the forest for ten miles just to being it home for the family.

That sounds like a great story. But, if we were to talk to Grandpa, we could learn the truth about that story and find out if it really happened or not and what actually happened to make him lose his leg. This way we could actually find out what happened and if it was a great story somebody made up or if it was the truth.

Chapter 10
Contacting the Spirit

As I mentioned earlier, I will be using three objects to contact a spirit. These are DNA, a photograph and an object from the person. Let's begin first with a small explanation of DNA.

DNA (Deoxyribonucleic acid) is a molecule that carries most of the genetic instructions used in the development, functioning and reproduction of all known living organisms and many viruses. It was first isolated by Friedrich Miescher in 1869.

Its molecular structure was identified by James Watson and Francis Crick in 1953. The study of DNA has only recently been researched on a major scale.

It has been used in many court cases to provide positive identification in murder cases, along with positive identification of family matters concerning the parents of children.

It has also recently been used to help genealogists verify their relationships with their ancestors. Police have used this technique to find the identities of bodies of murder victims found without anyway to know who they were.

The atoms in the strands of the DNA double helix consist of five gases the human body has. They are hydrogen, oxygen, nitrogen, carbon and phosphorus. These five gas elements are also listed in the previous chapter in the list of gases I was planning on using in my research to connect with the spirit.

The thought occurred to me, is it possible that spirits of the deceased was actually the DNA of the person? If this is correct, using a DNA sample of the deceased would attract the spirit to speak with.

I believe the DNA would be similar to a magnet, trying to reconnect with the main strand it was once connected with. There are no two people who have exactly the same DNA in their bodies. Even identical twins will not have exactly the same DNA as the other.

DNA could be considered the fingerprint of a human being, with no two being the same. It is known that even identical twins do not have exactly the same DNA or fingerprints, which means that every spirit would not be exactly the same as any other.

Theory – A persons DNA should remain connected with the person it belonged to, even in the spirit world. If the DNA sample is placed into the cabinet, along with the proper frequencies, that spirit should feel a pull to come and speak with us.

It would help the STEC to have not only the DNA, but the proper living point (L.P.) to be able to call them for a talk.

Now, let's talk about using a photograph as a contact for our spiritual guests to connect with. When I first came up with the idea of using a photograph of a person I wanted to speak with, it seemed simple at the time.

All I would have to do is to put the picture into the cabinet and their spirit would come and talk to me, right? Nope! I would later find out my research wouldn't be simple at all.

Once I became convinced the possibility of multi-dimensions was possible, I figured the moment the photograph was taken that it would solidify that decision. Once that happened, it would have created a permanent point of time when the new reality was created.

Using the photograph would become necessary to use as another contact to speak with the spirit, along with the LP for that time period.

When the camera was first invented, many members of the Native American tribes refused to allow anyone to take their pictures. They believed the machine would steal a part of their spirit.

Although it would not steal a part of their spirit, it would create a different reality where the spirit would continue existing. Taking the photograph would create a marked time when the decision was made, therefore time-stamping the point of the creation of the parallel reality.

If the scientific knowledge of today would have been available to the early Native American tribes, perhaps there would have been more photographs to record their history.

It would be interesting to find an original photograph and have the opportunity to speak with a member of the Native American tribe to learn the actual history and ways of life from their time period.

This thought caused a concern to enter my research. If I were able to contact a spirit of the past, such as the Native American spirit, how would I be able to speak with it?

I realized that if I wanted to speak with any spirit who spoke a different language, I would have to come up with a computer program that would instantly translate their language to the owner so a conversation could be had.

It is thought that there are around 7000 languages spoken around the world. This does not include the hundreds of ancient languages that are no longer being used.

Although the English language is spoken worldwide, it is still the ranked third in the list of spoken languages around the world. Mandarin and Spanish are the first and second most spoken languages. It has been noted, however, that the English language is the number one on the list for people who have learned a second language.

For the time being, I will be experimenting with the STEC and trying to speak to the spirits whose first language was English. It might take a bit of time to find the programs needed to translate all of the almost 7000 languages to be able to speak with any spirit I choose to have a conversation with.

The third way to contact a person spirit would be to use an object that was close to them. This could be anything they had during their lifetime that meant something to that person.

One of the most useful objects would be a ring. It has been said the vibration of a person can be absorbed by an object if held close to the body. Usually, a wedding ring is created and worn by only one person during their lifetime. Most people I know continue wearing a wedding ring, even when their spouse passed away.

The longer a ring is worn, the more of the owner's vibrations will be absorbed.

It has been noted that other materialistic items have been connected with spirits. According to my ghost hunting friend, there are many spirits who continue to stay around different objects, such as a special book, a particular picture, or even a vase that might have held flowers on their dining room table.

He continued to tell me that most of the time, the object was new when the person received it, and not connected with any other person. For example, it could be a new billfold, or a ring, or even another piece of jewelry other than a wedding ring.

Richard told me that his great-grandfather had a lucky coin he had carried with him for many years. I began to think about the possibility of something like that, but thought that the coin would have been held by many others, therefore absorbing some of their vibrations as well as his great-grandfathers.

When the time comes, I will be using a wallet my grandfather had. He used the wallet for many years, keeping it close to him constantly. I will also use the wedding band he wore from the day he married my grandmother, even after she was taken from him. Using these two items should allow me to speak with him using the STEC when it is built.

If I had each of the three objects needed, the chances of a successful discussion with the right spirit would increase. But, I feel that having one of the three will allow the STEC to work properly.

Chapter 11
The First Attempt

One rainy day, I decided to put the STEC together, just to make sure what I have designed so far would actually work. Starting early in the morning, I continued checking and double checking everything as I pieced the STEC for its maiden journey.

Not wanting to try to call anybody for a conversation yet, I didn't place anything inside the cabinet to attract a spirit. I sat down at the computer console, beginning to start up the programs I installed.

As the programs started, I reached over and flipped on the switch for the fog machine to fill the cabinet. I watched as the fog began to appear, watching carefully for any signs of leaking around the front glass door. I could see no sign of any leaks.

Satisfied, I turned back around to see the programs were ready to begin the testing with. One of the main tests I wanted to try today was to see if there would be any other noises coming from the computers or any other pieces of machinery I was using that could be heard coming out of the speakers.

I could tell right away that I would have to do something about the humming coming from the fog machine. If I could hear it, I knew the sensitive microphone would definitely pick up on it.

Since everything was running, I ran out to the garage to look for a large Styrofoam cooler I could use to cover the fog machine, hopefully to block out the noise. It took a few minutes to find it, but I knew immediately the cooler would do the trick.

Coming back to the room, I quickly noticed the sound of the fog machine could be heard clearly. I carved a hole in the cooler for the fog hose, turned it over and placed it over the machine, quieting the noise in the room.

I glanced over at the cabinet to see if there was any fog leaking out, not to see any. The fog continued to stay inside, swirling softly. Reaching over, I flipped the switch to turn off the fog machine from sending any more into the cabinet.

Satisfied with the tests so far, I turned on the microphone inside the cabinet to see if I could hear anything coming out of the speakers across the room. Walking towards the STEC, I tapped softly on the front glass door, hearing the tapping coming out of the speakers. I knew that the microphone was working properly.

Going back to the computer, I turned off the overhead lights in the room and turned on the ultra-indigo lights of the STEC. The swirling of the fog inside the cabinet had slowed enough to allow just a haze to remain.

Thinking this would take a while to make sure every aspect of the STEC was working, I went out to the kitchen for a bottle of water. My kitchen is only about thirty-feet away from the STEC room, so I knew it would only take a minute or so to get the water and return.

As I was about to return, I could hear a faint voice coming from the other room.

"Where is he?" I heard female voice say softly. I hurried back to the room, looking at the cabinet, noticing the fog was no longer still, but was swirling slightly.

"Are you still there?" I asked, watching the fog beginning to settle down. I waited a few minutes before walking back to the computer console to replay the recording while I was away.

I listened to the recording for the previous five minutes, hoping I wasn't just hearing something that wasn't really there. To my delight, I could hear the voice once more, unfortunately while I was in the kitchen getting the water.

I was becoming confused, knowing I had most of the programs running but I didn't have anything placed in the cabinet to call a spirit to talk with. I looked at the programs I didn't have running, realizing I didn't have two of the necessary programs going. The first was the sound speed inside of the cabinet and the second was, what I thought to be the most important, was the calculations to find the point of time of a spirit.

Stopping the video recorder, I rewound it back to the same time I heard the audio part. I watched as the fog inside the cabinet began to swirl, taking the form of a female spirit. The spirit seemed to look around the room before disappearing just as quickly as she showed herself.

I looked at the spirit in the video, not recognizing any resemblance to a person I might have known. Many thoughts crossed my mind as I tried to figure out who the spirit could have been. Is it possible this spirit could have been what people would call their guardian angel? If so, I sure hope she would show up again, because there are quite a few questions I would like to ask her.

I never thought of the possibility of using the STEC to converse with my own guardian angels. Since they are always around me, would it be possible for them to speak to me so I could understand what they are telling me? This could open a new door to the way I would be able to live my life.

Sitting back in my chair, I began to think of the possibility of speaking with my invisible helpers on this earth. The possibility to actually see them and speak to them in person excited me. Although I have tried to live a good life, knowing I could actually talk with them would allow me to create an even better and more correct lifestyle.

Everyone has a guardian angel, but like most, it is often impossible to understand the ways they show you how to live your life. If it were possible to have a talk with them in the mornings before you get your day started, imagine how much better your life could be.

Sitting in a chair facing the cabinet, I was hoping the spirit would once again show herself to me. As I waited, the thoughts of what type of questions would I ask and what questions would she be able to answer entered my head.

It is probably a good thing she didn't come back right away, because I couldn't think of too many questions I thought she could answer. Quickly, I made a note on a piece of paper to remind me to research this possibility a little more.

Satisfied with the way everything was working, I walked back to the console and began to shut down the STEC until I had more time to concentrate on its performance. The first thing I had to do was to turn on the exhaust fan inside the STEC to remove the fog, still swirling slightly in the cabinet.

I had connected a small exhaust hose to the bottom of the cabinet which led to the outside. I did this for one main reason, so the fog would not stay in the STEC room before disappearing. I'm sure if anybody could see my house, they would probably call the fire department, thinking my house was on fire. After turning on the exhaust fan, I looked outside, watching the fog escaping being taken away with the breezes, just the way I planned it.

Chapter 12
My Guardian Angel

I have always been a big believer in guardian angels. I was taught at a very young age that they exist and help us in our everyday lives.

Even as an adult, I still believe in them. There have been many times when I knew it wasn't me making the choices, but somebody, or something, was always there to help me along the way.

Thinking about the female spirit I heard and saw in the cabinet the other day made me wonder if that could have been my own guardian angel coming to pay me a visit. If it truly was her, how would I be able to contact her using the three methods I would normally use?

Obviously, I don't have a picture of her, or any of her DNA or a personal item she once had that I could use to contact her with. I walked back into the STEC room, deciding I was going to attempt to talk with her.

I turned on the computers and the programs I would need to run the STEC with, still trying to think of a way to ask that same spirit to come in and talk with me. I have always had the habit of talking out loud whenever I am alone or needed to think out a problem I was working on. Perhaps this would be the way to ask for the spirit to come and speak with me.

Knowing I didn't want to miss my chance to speak with the spirit as I did before, I walked into the kitchen and took a couple bottles of water out of the refrigerator, just in case it was going to be a while before she would be able to show herself. I walked back into the STEC room, closed the door behind me and sat down at the console to make sure everything would be ready for her arrival.

After checking the sound system and video leads, I was happy with the results. The last thing I had to do now was to turn on the fog machine which would make it easier for me to see the spirit if she could come and talk with me.

Soon the fog began to fill the machine, swirling slowly in the cabinet. Usually it takes about five minutes to fill the cabinet and begin to still itself. After turning off the overhead lights, I sat down in the chair in front of the cabinet, still calling out into the air hoping the same spirit that was here a few days ago would stop by for a visit.

I have to admit, it was almost hypnotic watching the ultra-indigo lights playing with the fog as it moved slightly inside of the cabinet. Soon, my eyes were beginning to see things that weren't actually there. I guess I was hoping that the swirls would begin to take the form of a spirit.

Ten minutes passed before I began to see the fog mysteriously begin to move faster inside the airtight cabinet. Not only would this prove to me the STEC was working the way I hoped, but this was also the very first spirit I would have the pleasure of seeing and speaking with.

I don't think there was any way I could have prepared myself for the sight I was about to see. After twenty years of working on the STEC, my dreams were finally about to come true.

I watched as the swirling of the fog began to stop, allowing me to look upon the first spirit I would have a chance to speak with. At first glance, I was wondering if this was a true angel, realizing that it could be since I had never seen one to begin with.

I stood up from my chair as the spirit continued to show herself to me. In another minute I found myself face to face with an angel of a woman I would estimate to be around forty years old. Although the colors weren't as sharp as I would like, I could make out every detail of this young woman.

I can only describe her facial features as your typical girl next door looks. Black hair cascaded over her shoulders, almost touching her waist and gently waving in the fog as if there were a gentle breeze in the cabinet.

Her long gown reached from her neck to the floor where it disappeared because of the fog settling in the bottom of the cabinet. Her eyes were bright and cheerful but did not compare to her smile when she saw me looking at her.

"Hi, James," she said, letting me hear her voice through the speakers in the room.

"Hello," I said, suddenly becoming at a loss for words to say. Even though I had many questions in my head I wanted to ask her, my mind became completely blank.

"Can you hear me alright?" she asked, looking strangely at me.

"Yes. Are you able to hear me?"

"I can hear you just fine, James," she said, beginning to smile again. "I would like to be the first to congratulate you on completing your invention. I am honored to be the first to come and speak with you."

"It is my honor to be able to speak with you," I told her, finally coming back to my senses. "Is there a name I may use to talk with you?"

"You may call me Marie, if you would like," she said, keeping her hands folded tightly in front of her. "Please, James, sit down if you want to."

"Thank you," I said, sitting back down in my chair but not taking my eyes off of Marie. "Are you my guardian angel?"

"I have been asked to come and help you during this time of your life," she told me. "I am one of many who have been helping you throughout your lifetime."

"I realize there are many subjects I can't ask you about," I said, trying to figure out where the boundaries of my questions would be, "But can you tell me if I am doing alright the way I am living my life?"

"You are doing just fine," she told me, smiling back at me. "There are a few things we will be working on, but overall, you are doing fine."

"If I have a problem or decision to make," I thought, not sure of what her answer would be, "Would it be possible for you and I to talk like this so I can figure out the right way to do things?"

"I am sure we can meet like this once in a while, but not for everything you must go through," she said, trying not to tell me exactly what I was hoping to hear from her.

We continued talking for a while about different parts of my private life before she turned the conversation back to the STEC invention.

"There is one major part of this machine that you must address," Marie told me. "As you know, there are both good and evil spirits. Although you have been able to break the barriers between the two worlds, there is the possibility an unwanted spirit to come and speak with you. There are many different ways to not allow this to happen, so I believe you should add that security to the STEC."

"Would you help me with that?" I asked Marie.

"We will all be helping you," she said, making me understand the importance of listening to the angels who are around me daily.

"Is it comfortable in the cabinet for you, or is there anything I should change to make it better?"

"The temperature is fine," she began, telling me what a spirit would be feeling while inside the cabinet. "One of the main problems I see right now is the volume of the speakers you have installed in the base of this cabinet. Perhaps if you could lower the volume of the bass, it would be easier to hear what you are saying."

Marie and I continued talking about what I could do to make the spirit more welcomed when they came to speak with a person. I thanked her many times for helping me with the things about the inside of the cabinet that I have not been able to experience and feel.

Soon, it would be time for Marie to leave me. She assured me she would be able to come back and talk with me again when she felt it was necessary. I could not thank her enough for everything she has helped me with, not only with the help with the STEC but with everything else that her and my other guardian angels have been helping me with during my lifetime.

I said good-bye to Marie, leaving me with a feeling that our bond had just become much stronger.

Chapter 13
Debbie

Alright, we now have everything ready to have the STEC contact the next spirit I want to talk with. We know it is possible to speak with a spirit more than once, as long as I write down the vibration to dial in on. It was a great honor and pleasure to speak with Marie, letting me learn more than I ever thought I could from her. Now is the time for me to try to contact the spirit that made me think about building the STEC.

Not having any of Debbie's DNA or personal items, I carefully placed a picture of her inside the machine, hoping to draw her spirit into the fog to let me talk with her once more.

At the time, I wasn't sure if I was more nervous about the machine working the way it should, or actually having the chance to talk with Debbie again. I turned off the lights in the room, excited to finally talk with my friend who passed away many years ago.

Soon, the ultra-indigo lights began to create an eerie feeling in the room which intensified as the fog began to fill the STEC. Looking over at the computer screen, I was satisfied when I saw the signs for the recording and speakers were on, letting me know this conversation was ready to be recorded.

Sitting down in my chair in front of the STEC, I waited in anticipation for a spirit to show itself. The fog was swirling in the cabinet, causing me to become more excited about seeing my friend once more. From the center of the fog, I could see a small void beginning to appear.

Soon the fog would allow a spiritual form to be seen more clearly. Even though the spirit was a grayish color, I could still recognize it as being my friend, Debbie. It took a few seconds for her form to become comfortable within the confines of the STEC, but I could tell she was as happy as I was to finally be able to talk to each other.

"Can you hear me?" she asked, using the voice I had heard for many years and have missed terribly for the last twenty years.

"Yes, I can," I answered her, with excitement in my voice. "Can you hear me alright?"

"Yes," she said, beginning to smile at me.

"May I ask you a couple of questions before we talk?" I asked, wanting to make sure this was indeed the spirit of Debbie.

"Yes."

"Do you remember when we met down here on earth?" I asked.

"We first met when my parents moved to town and I started at the same school you went to," she answered, knowing I was testing her to make sure she was the right spirit.

"When we got married, who was your maid of honor?" I asked, trying to keep a straight face.

"We never got married," she said, smiling back at me. "But, I wish you would have asked me. I would have said yes."

I was now convinced that this was the true spirit of Debbie. I chuckled at her last answer, letting her know I was still kicking myself for never asking her.

We continued making small talk for a while before she said something that made me question a few of my thoughts about the afterlife.

"Did you know Sam died?" I asked, wondering if she knew or not.

"Yes," she said. "I was with him when the time came. I had watched over him, just like I watch over you sometimes. I was with him when his spirit left his body."

"Have you been able to see him since he left here?" I wondered, hoping she might give me a small glimpse into the afterlife.

"When he finished his life on earth," she began to explain, "all of his friends and family were waiting to see him again. We all welcomed him with opened arms, letting him know we were there for him. I have been able to see him a few times since then."

"I will be trying to contact him soon," I told her, glad to know my two best friends were able to see each other again.

"I am sure he would be happy to talk with you again also."

I could tell by the way she started acting that there was something else happening.

"I am going to have to leave soon," she said, in the voice I never got tired of listening to. "There is someplace I must be."

"You have a job up there?" I asked, wondering what she was talking about.

"It is not really a job," she explained, "But it is more like being a helper. I have a few people I help while they are living on earth. Just like you have your helpers, I help those who need it too."

"Are you one of my helpers?" I asked, hoping she would say yes.

"No, but I have been around to see you a few times," she said, looking at me with her beautiful smile. "I am so proud of you, Jim.

You have always been in my heart, just as I have been in yours."

"Can I see you again sometime?" I asked, hoping she would say yes.

"Of course you can. You have my number, don't you?" she asked, chuckling.

"I do now."

"I have to leave you," she said, beginning to disappear into the fog. "I have been waiting for this moment for a long time. I am glad you were able to accomplish this great feat. I love you, Jim."

"I love you too, Deb," I said, feeling a small lump forming in my throat. I watched as the fog in the STEC began to swirl slowly, knowing my time with Debbie was now finished.

I never would have imagined the different feelings I would have after being able to talk with Debbie. The emotions that were flooding my thoughts were almost euphoric.

Chapter 14
Grandpa

Although the STEC has been working as it should, there was still one spirit I wanted to speak with. I'm not sure if I was too nervous to try to contact him or not wanting to know how he feels about the way I have turned out since his passing.

Grandpa has always been in my heart and I think about him every time I look at the old birdhouse he and I built almost thirty-five years ago. I finally decided to put my fears aside and try to locate the spirit of my grandfather so we could have the talk I have wanted for such a long time.

The room became eerie once again as I turned on the ultra-indigo lights and shut off the others. The signs on the computer monitor let me know everything thing was in place for me to get hold of Grandpa. I entered the information needed, letting the computer do its job of making sure everything was ready before I began.

The final piece of the connection would be a ring my grandfather wore throughout his lifetime. The DNA on the ring would work as the connection to call Grandpa from the other realm so we could have our talk.

After placing the ring in the machine, I reached over and turned on the fog machine, filling the STEC with a thick fog, making it easier to see Grandpa if he was able to pay me a visit.

Quite a few minutes passed, making me wonder if I properly set up the STEC for Grandpa's arrival. I was just about to get up to check when I saw the fog beginning to swirl. I sat back down, watching with anticipation for the spirit to show itself.

I tried hard not to smile when the spirit form started becoming clear. There was no doubt in my mind that I was finally going to be face to face with my grandpa.

As the spirit form became clear, I could finally see my grandpa smiling at me.

"Hi, Jim," he said, becoming as excited as I was to see each other after thirty-five years.

"Hi, Grandpa," I answered him, wiping a couple of tears from my eyes. "How are you?"

"I am one proud grandpa," he said, still smiling at me.

"I never forgot those three words you told me while we were building that doghouse in Montana, Grandpa," I said, hoping he would correct me on my statement.

"What doghouse?" he asked. "I thought we built a birdhouse, not a doghouse."

"That's right," I told him, knowing now this was in fact my grandfather. "Do you remember those three words you told me never to forget?"

"Do you mean," he said, with his smile becoming bigger, "Conceive it, Believe it and Achieve it?"

"Yep, those are the right ones," I said, smiling just as much as he was.

"It looks like you really listened to me, boy," he said, looking directly at me. "I am so very proud of you. So are the others. Your mother and father have been watching over you and are as proud as your grandmother and I are."

"Thank you. If you didn't teach me those three words while we built that birdhouse, I don't think I ever could have succeeded in this."

"Sometimes the smallest conversations have the largest effect on people," Grandpa told me, with a more serious tone in his voice. "I am just glad you were listening to me."

"I have never forgotten those words," I told him, knowing if it weren't for him, there would be no way I would have been able to succeed in building the STEC.

"You have been living your life by those words, Jim," Grandpa said, still looking at me. "Look at everything you have been able to do.

You have been a success throughout your life, yet still remained faithful and helpful to everybody you know. That is the sign of a great man. It doesn't matter if you invented this machine or not, we will always be proud of you."

I could feel the lump in my throat begin to reappear, making me softly thank Grandpa for his words.

We talked for a while longer with Grandpa telling me how he and Grandma were, and how my parents were doing. Each of them was busy, just like Debbie was, helping the living people on the earth.

"Can you tell me what it's like up there?" I asked, not sure if he could tell me or not.

"There are some things I can tell you," he answered me, "But there are many I cannot."

"Are my thoughts about parallel lives correct?" I asked, not knowing what kind of questions he could answer.

"Yes, they are," he said, surprising me. "When a parallel life is created, a part of a person's spirit is placed there. After the death of each part of the spirit is complete, it is at that time it will be called to be with the Lord."

"So, whenever a piece of the spirit passes away on a different life," I asked, "does it join up with the other pieces of that spirit?"

"Yes. When the spirit is complete, it will be called to meet with the Lord and be judged."

"Have you been with the Lord yet?" I asked, wondering if all of his parallel lives spirit has become one yet.

"Not yet."

"But you passed away so long ago," I said, confused about why his spirit wasn't whole yet.

"Just because I passed away at a certain age here, doesn't mean I passed at the same age somewhere else. When I passed in this life, my spirit joined with the other parts that passed before I did. Does that make sense to you?"

"Yes, it does," I told him, soaking in everything he was telling me.

We continued talking for a little while longer before he said he had to leave. I was going to ask if I could call on him again, but he surprised me when he asked if he could come back and talk to me again. I didn't have to think it over and told him I couldn't wait until the next time.

Chapter 15
A Call to Talk with Thomas

(Please remember that this is a fictional trip and discussion with Mr. Edison.)

I'm not sure how many times I went over everything, making sure the STEC was ready for its first time. Throughout the tests I've done, I felt comfortable the STEC was going to work this time. The question is, who am I going to try to contact next?

I remember thinking about the possibility of only being able to speak with a spirit just once. The first spirit I spoke with was my guardian angel, Marie. After that I had the pleasure of speaking with Grandpa, Debbie and Sam again. I found out from speaking with them that I would be able to speak with them again and not have to worry about just having a one time talk with the spirits.

This decision would be made for me, although I was thinking along the same line in my own thoughts.

One night, I was going over a few of the calculations I made, trying to tune up the machine even before turning it on for the first time. I had to put the papers down when my telephone started ringing in the den. I don't have a telephone in the STEC room, not wanting it to interrupt my time speaking with the spirits if it began ringing during a session.

A few months ago, I began a page describing the STEC on one of those community websites. Since that time I have acquired a small following of people interested in the STEC.

I am bringing this up at this time because of the telephone call I received. One of the followers of my web-page is involved with one of the Edison facilities in the United States.

They had just called and invited me to bring the STEC to their location, wanting to see if it would be possible to speak with Thomas Edison once more. I considered this to be a true honor and quickly told them I would visit with them.

After all, who would be the best spirit to try to contact other than the man who first came up with the idea? After talking with them for a couple of months, during the last call it was decided that they would be sending down a truck in one week to help me transport the STEC out east to make sure it would arrive there safely.

I spent the next five days breaking down the STEC and boxing up the different components, making sure I would not be leaving anything vital behind. I looked at the room, surprised at the number of boxes lined up against the far wall, each with the names of the different components listed on the outside.

Now came the hard part. How was I going to pack the cabinet itself? After figuring it out, I left the packaged cabinet on the floor, knowing it would be safe until the movers came in a couple of days to take it from the house.

Finally, the day came to begin the journey to the east coast. I woke up about four-thirty and began waiting for the moving truck to pull into my driveway. I was told they would be there between six and seven o'clock in the morning to begin the move, so I thought I might have a few extra minutes to double-check my list of things to take with me. Curious, I walked out on the front deck, just to see the movers standing by their truck already parked in the driveway, waiting to get their day started.

The two men took me up on my invitation to come inside and enjoy a cup of coffee while I got dressed to start the day. After a quick shower, I walked back out and started talking with the movers about what was about to happen.

The taller of the men stood and gave me a round-trip ticket I could use to travel out east while the STEC was making its way there on the ground. They also told me there was a limousine that would be here in a few hours to take me to the airport so I wouldn't have to drive.

"How am I supposed to get home when I come back?" I joked with them. I was told it had already been planned that the limo would be there on my return to bring me home again. The STEC would be arriving back home two days after the testing out east was finished.

After our second cup of coffee, I led the two men to the back room that held the boxed parts of the STEC. They were both surprised to see the boxes stacked neatly and labeled with what was inside. I offered to help them move the boxes, but they told me they were already prepared and had their plans for stacking them inside the moving truck.

I walked outside with a small box, curious to see the back of the truck. The sun was still down, not expecting to rise for about another hour or so. One of the men went out and turned on the lights in the back of the truck, making it shine brightly in the driveway.

The inside was as clean as the outside of the truck. All of the walls were covered with thick padding. Many hooks lined the walls, letting me know my boxes were going to be strapped in tightly with some of the many nylon straps rolled up neatly in the far left corner.

They placed the small box I was carrying to the side, bringing out the STEC cabinet first.

The men unfolded a large blanket, placing it on the floor of the truck. I just stood back and watched these two as they continued packing each of the boxes around the cabinet, putting a layer of padding between each box.

By the time they finished, I was comfortable in the fact that the STEC was going to have a safe trip, arriving just as it had left here. I thanked the two movers, watching as they pulled away for the two day trip to the east coast.

Walking back into the house, I got the feeling I had been wanting for a long time. The feeling of possible success was beginning to show itself. This was the big break I had been praying for. I was now at the point of no return with the STEC.

Chapter 16
The Set-up

I've have never been one who enjoys flying, but knowing this was a great opportunity for me, I gathered the strength to board the plane. As the plane lifted off the ground, I was hoping I would still be on this side of the STEC as a human being, rather than becoming a spirit coming in from the other side to talk to somebody else.

The pilots of the large jet successfully landed safely, letting me take a deep breath and thanking the good Lord above for not letting me die today. I could now focus on the STEC, getting it ready to make its debut in front of many interested people of the Edison companies.

Getting my suitcase from the carousel, I started walking towards the front exit to find a cab to take me to the hotel where I had my reservation at. I started feeling important when I saw a man holding up a sign with my name written on it with big bold letters.

I introduced myself to the man as he showed me the way to another limousine my hosts had sent for me. There was a strange feeling when many other travelers began staring at me when the limo driver opened the back door of the limousine for me to get in.

Joseph, the limo driver, told me he was to take me to the hotel I would be staying at where I would receive an itinerary of what I would be doing for the next few days during my stay. Joseph and I continued talking as he drove to the hotel, with him telling me he was assigned to be my chauffeur during my stay.

Enjoying the scenery of the east coast, I was happy knowing I wasn't the one who was doing the driving through this traffic. Back home if I saw more than two cars in a row, it meant there was a parade going through town.

Joseph and I continued talking as he drove me to my hotel. He pulled up to the main doors, drawing the attention of many of the others in the area. Getting out, he straightened his black jacket and put his hat on straight.

Once again, I felt like a famous person getting out of the back of the limo, as many pairs of eyes suddenly began focusing on me. Joseph held the car door open for me, closing it as soon as I stepped out. I walked inside the hotel as my driver opened the trunk, taking out my suitcase before following me inside to the registration desk.

"Here is my cell-phone number," he said, handing me a business card. "Call me if there is anywhere you want to go. If I don't hear from you tonight, I will be here at eight tomorrow morning to take you to your first meeting."

"Thank you, Joseph," I said, reaching over to shake his hand. "I'll see you in the morning."

I turned my attention to the clerk behind the counter, waiting to complete my registration. Everything was already set for me, leaving the only thing for me to do was sign my name one time.

I was given the electronic key to a room on the seventh floor of the hotel and a sealed manila envelope with my name on it. Suddenly, I felt more like a spy instead of an inventor.

After making it to my room, I looked out the window to admire the view of the town. I was sure I would enjoy the time here, but was already looking forward to the slower pace of living I was use to back home.

The following morning, I met Joseph in the lobby of the hotel and followed him to the limo waiting to take me to my first meeting of the day.

Joseph dropped me off at the front doors of a large stone building, telling me the receptionist at the front deck could help me if I had any questions. I thanked him, letting him know I would be ready at four when he came back to get me again.

The kind receptionist gave me a visitors badge and directions to a room that held my transported STEC parts. She was kind enough to let know my host, Mr. Johnson, would meet me there in a few minutes. I thanked her and started walking towards the room I would be allowed to rebuild the STEC in.

I smiled when I looked at the number of the room number 12. *(I will explain the importance of that room number later.)* I was surprised when I walked into the room, looking at all of the boxes placed neatly throughout the room. Lying in the middle of the room was the largest and most important box, the cabinet.

I looked around the room, noticing it was exactly the type of room I asked for when I agreed to come here for the demonstration. There were no windows to allow any light through and a small sealed office where I could keep the electronic parts of the STEC to make sure no outside noises could be heard.

I was just about to inspect the boxes for any damage when I heard my host enter the room.

"Mr. Steines?" he asked, reaching out his hand to me. "I'm Larry Johnson. It is a great pleasure meeting you."

"Mr. Johnson," I said, shaking his hand, "It is all my pleasure. Thank you for inviting me today. Please, call me Jim."

"As long as you call me Larry," he said, releasing his hand from me, chuckling slightly.

"I hope this room will serve your needs while you are here with us."

I told him that it looked great and asked him how long I would have to set it up before the demonstration would take place. He told me I would have the rest of the day and tomorrow to get things ready, but first, there was a small group of people who would like to meet me before I started setting up the STEC.

We walked out to the hall, making small talk as he led the way to another room a few doors down. We entered to find about ten others waiting for our arrival.

Larry introduced me to the people interested in meeting me. I was honored to see my work was welcomed by so many others. Although we were just speaking for a few minutes, I was becoming anxious to go back to my room and start piecing the STEC together for the demonstration the day after tomorrow.

We continued talking and getting to know each other for about an hour before leaving each other for the day. I found room 12 again, and started to unpack the many boxes to put the STEC back together again.

Chapter 17
The STEC Presentation

The next day, I was just putting the finishing touches on the STEC when Larry walked through the doorway. I smiled when he looked at the STEC almost ready for the demonstration.

"This is truly quite an accomplishment, Jim," he said, walking over to take a closer look at the viewing cabinet. "I can't wait to watch the demonstration tonight."

"Thank you," I said, looking over at the wires on the floor, hoping I connected them to the right machines. "Is there a convenience store close by?"

"Yes, there is. Why?" he asked, looking strangely at me.

"There are a couple of things I left back home that I might need. Do you think there would be enough time for me to go over there before the demonstration today?"

"I can send someone over for you, if you would like," he said, taking out a small notepad from his jacket pocket. "What do you need?"

"If it wouldn't be too much of a problem," I said, looking over the controls, "I need a large Styrofoam cooler, a roll of duct tape and a three musketeer's candy bar."

He looked strangely at me as he wrote down what I needed. I could tell by the strange smile on his face that he thought he was talking with a true southern redneck.

"Is that all you need?" he asked, trying hard not to chuckle at my request.

"That should be about it," I told him, looking around to make sure there was nothing else I left back home.

Larry assured me the items I asked for would be in the room waiting for me after the talk I was asked to give to a few invited guests. Putting on my tie, I looked around once more, making sure there was nothing else I would need for the day.

Although I am not one who likes to speak in front of a group of people, I gathered my nerves and walked with Larry to the room where I would be giving my short speech and answer questions to a few of the invited guests.

As we rounded a corner of the hallway, I quickly realized there would be more then just a few people. We were both welcomed with bright flashes of cameras as we walked around the corner to the hall leading to the room. There were also a couple reporters and cameramen from some of the local television stations waiting to record our every move.

I followed Larry towards the front of the room, where he stood in front of the large crowd to introduce me. Since this was the first time for me to speak in front of a group of people, I didn't have any note cards with me to refer to. This would be the first and last time I wouldn't have notes to use.

After a very kind introduction, I walked up to the podium and shook Larry's hand. As I said earlier, I was not accustomed to public speaking, which the sudden nervousness reminded me of. I looked out at the large group, momentarily losing my train of thought about the STEC. I quickly gathered myself and began the presentation.

"Thank you, Larry, for the kind introduction," I began, still trying to overcome my fear of talking in front of a large group of people. "I will ask you all now for your patience while I talk with you. If it seems like I'm nervous, it's because I am."

The large crowd chuckled as I looked around the room at the number of people who came out to see me.

"Today, I am going to talk a little bit about my invention that I call 'STEC'. STEC stands for the Spiritual Tele-communication Center. This invention will allow a person to talk with a person who has passed away.

Quite a few years ago, I lost two of the best friends a man could ever have. For many years, I wished I could have talked with them again. One night, the idea came to me about building a machine that would allow me to call for their spirits so I could have that conversation with them once again."

"That night was over twenty years ago," I continued. "It hasn't been until recently that I was able to build the STEC. There were many set-backs, but I was able to finally build a working machine."

I was stopped suddenly by a small round of applause for the completion of the STEC.

"When I first started this project," I continued with my explanation of how I built the STEC, "I was completely lost of how I would accomplish this. I graduated high school over thirty-years ago, never thinking I would have to have a college degree in physics or any other higher mathematics. There were many times I wish I had continued to further my education.

There are three men who I must thank for allowing me to be able to accomplish this invention. The first is Charles Babbage.

Charles Babbage was an English mechanical engineer who is considered the 'father of the computer'. This man is credited for inventing the first mechanical computer in the early 1800's.

The other two are Sir Timothy John Berners-Lee from England and Robert Cailliau from Belgium. These are the two men who created the World Wide Web back in 1989.

If it weren't for these three men, I don't believe I ever could have found all of the information I needed to create the STEC. So, to the three of them, I give my undying thanks for allowing me to be able to search the web for the answers I needed.

The web made it much easier for me to find the answers to the questions I had about the different aspects of building the STEC. One of the most important pieces of information I found on the web happened a few years after I began this project.

I learned of another man who had the same idea that I was working with. His name was Thomas Alva Edison. Knowing this genius was also working on the same type of machine allowed me to believe the reality of the STEC might be possible.

Without letting you all know the mathematical equations and other parts of my research, let me tell you basically how this machine works."

Chapter 18
Presentation Conclusion

I reached over and took a drink of water from the bottle that I brought with me. Glancing at the audience, I could see a few of them getting out their notepads in hopes I would inadvertently be giving away a few of the secrets of the STEC.

"Growing up, I was taken to church with my parents," I started, presenting the second half of the presentation. "As a young child, I found religion to be confusing at times. As I grew and began learning about life, there was one more statement which confused me even more. I had heard that science and religion don't mix. After creating the STEC, I have to disagree with that statement.

As I began to research the different aspects of the STEC, I found myself looking in the strangest places for information. This would include almost everything from dogs and cats to quantum physics string theory."

I looked over the crowd, noticing that everybody was still with me as I began the explanation of putting everything together, therefore creating the STEC.

"There are many components to the STEC," I continued, realizing everyone in the room was listening to what I was saying. "I thought about the five sense humans have. Sight, smell, taste, hearing and feel. Which of these senses would we need to be able to have a conversation with a spirit?

The two obvious ones would be sight to see them and hearing to listen to them during the conversation. I'm not saying you don't need the others, smell, feel and taste, but they aren't as necessary as the others. So, I began trying to figure out the senses of sight and hearing.

I would find that figuring out the ways to hear and see the spirits would be easy compared to finding out what plane the spirits existed on. Finding out the location would end up taking most of the research time of the STEC.

After speaking with different people who go out and find spirit voices, or can photograph the spirits, I began to think about the location a spirit is removed to after leaving the body. If we are able to hear them and see them, then their plane of existence must be relatively close to our own.

When a person dies, their spirit becomes disconnected from the body. Even if it took only one-tenth of one second to disconnect, that would be enough to lose that amount of time, therefore allowing the spirit to become a part of the other plane of existence instead of ours.

It would be like becoming out of phase with our reality, yet still being able to see us but not talk with us. In the research I have done, the different voices from the past that I have been lucky enough to record all have one thing in common. They were all out of speed with our time. Some are faster and some are much slower than our own speed of talking. This led me to believe that those spirits who were speaking with me are all at a slightly different time wave, depending on the year they died."

I could tell by looking out into the crowd, many of them were beginning to get confused at my explanation.

"Knowing there would be a difference in the speed of their voices," I continued, trying to simplify my talk, "I came up with a voice program that would automatically slow or speed up the speech of the spirit I was speaking with, so we could carry on a real-time conversation."

I looked back out to the crowd, watching a few of their heads nodding, letting me know they understood what I was trying to tell them. Taking another drink of water, I continued.

"In conclusion, there are many applications to the STEC. To learn the truth about history, now we can actually talk to somebody who lived it. For me personally, I just built the STEC to be able to talk with my friends and family who have died. Not only is it a great closure to be able to say good-bye to them, but it's also a great opportunity to speak with them before I am called to the heavens when my time on this earth is finished.

I would like to take this time to thank this great organization for giving me the chance to come and speak with you all today. This afternoon I will be using the STEC to have a conversation with Thomas Edison. I'm sure you will be able to read about it on the internet when it is finished.

"I have time for a couple of questions," I said, looking over at Larry, standing against the far wall, shaking his head yes that I would have time. I looked out over the number of people with the hands raised and pointed to a young lady, about fifteen or sixteen years old, sitting a few rows back from the front. "Hi, young lady. What's your name?"

"Julianna Kirby," she said, standing up so I could see her better.

"Hi, Julianna. It's nice to meet you. Do you have a question for me?"

"I have to give a report for my science class tomorrow," she said, looking right at me. "Could I interview you for the report after you are finished here?"

I took out my notepad and jotted down a quick note before I answered her question. Taking a look over at Larry pointing to his watch, I knew he was telling me I wouldn't have much time for an interview with her, but I knew it would be important for Julianna if I said yes.

"I would be honored to talk with you," I told her, watching her smile.

She sat back down as more hands began to raise for the question part of the presentation.

"How about the gentleman in the back?" I asked, looking around the room at the hands in the air.

A rather burly man stood tall over the crowd to ask his question.

"First of all, congratulations on your success with your invention," he said, giving me a thumbs up for my accomplishment. "What other uses would there be for the STEC besides talking with your family or famous people?"

"There are many other applications that can be used with the STEC," I said. "The police department could use the STEC for identifying murders, for instance. Let's say the police are working on a murder case and they have no solid leads. Having the STEC in their department could be used like this.

They take a DNA sample from the victim and place it in the STEC. In a few minutes, the spirit of the murdered victim would appear and let the police know who killed them. Now the police have a solid lead in the case.

Another use for the STEC could be at a funeral home. Imagine going to a funeral to say good-bye to a loved one or a dear friend. If the STEC were to be used, the departed person could also say good-bye to you.

Let's say you are really interested in history, and your favorite person was Abraham Lincoln. Using the STEC, it would be possible to actually have a talk with him, as long as you have the proper items needed to contact him with."

I looked over at Larry as he started moving closer to the podium, letting me know it was about time to stop.

"I want to thank you all for taking time out of your busy schedules to come out and see me today," I said, beginning to close out my talk. I looked over and handed Larry the piece of notebook paper I scribbled on, watching him smile and shake his head yes.

"A couple of minutes ago, Miss Julianna asked me if she could interview me. Now, I have a question for her. Julianna, are you afraid of ghosts?" I asked, listening to a little bit of chuckling from the rest of the audience.

"No," she said, wondering why I was asking her.

"Would you do me the honor of joining me at the demonstration this afternoon so you can see how the STEC works while we have a talk with Thomas Edison?" I asked, watching her look over at her dad sitting next to her. He shook his head yes, letting her know it was her choice.

"Can Dad come along too?" she asked.

"Sure, as long as he isn't afraid of ghosts," I told her, chuckling with the others in the room. She looked over at him as he assured her that he wasn't afraid of ghosts.

Ladies and gentlemen," I concluded, "Thank you all for coming out today. Let me leave you all with one thought. If you had a chance to use the STEC, who would you talk to?"

After the applause stopped, Larry stood in front of the crowd, letting them know I had my books for sale outside in the hall and that I would stay around to sign them, if they wanted. Soon, he joined me as we walked outside to the table he had set up for the book-signing.

I excused myself for a minute when I saw Julianna and her father walking out of the room. I reached over, taking one of my books off the table and signed it for Julianna. She smiled when I gave it to her, telling her it was a thank-you for joining us this afternoon at the demonstration.

We talked for a few minutes before they left for a couple of hours until they came back about three o'clock for an explanation for the STEC before we started. I walked back to the table, joining Larry again to begin the book-signings.

Chapter 19
Getting Ready with Julianna

About an hour later, Larry walked me back to room 12 so I could put the finishing touches on the STEC for the final presentation. We walked in the room to find everything was just as we left it, with the exception of the addition of the Styrofoam cooler, a roll of duct tape and a large three musketeer candy bar I needed.

"It looks like Joseph took care of you," Larry said, handing me a small note resting on top of the cooler.

'Good luck, Mr. Steines,' it said, making me smile at Joseph's kindness.

"I have to ask you, Jim," Larry said, looking over at me, as I put the note back on the table. "I think I can understand why you need the cooler and duct tape, but why the candy bar?"

"The microphone is so sensitive," I explained as he began to smile, "I don't want it to record my stomach growling."

"I have to leave for a while," he said, shaking his head and chuckling. "I'll be back in about an hour or so. When Julianna and her father show up, would you like me to bring them in here for you?"

"That would be great, thank you."

For the next couple of hours, I tested all of the different elements of the STEC, making sure they were all working properly. I was just finishing up covering the fog machine with the Styrofoam cooler, taping around the base and the floor to keep the humming sound silent when the door opened once more.

I turned to watch Larry bringing in Julianna and her father, Tom. I could tell by the look on her face she was excited to have this chance to watch the demonstration with me.

"Thank you again, Mr. Steines, for letting me do this interview with you and being able to watch the demonstration," Julianna said, stretching out her hand to me.

"It is my pleasure," I said, shaking her hand and admiring the professionalism she was showing. "Thank you for bringing her today, Tom. Please, call me Jim," I said, shaking her father's hand as well.

"I don't think I would have had a choice," he joked.

"Julianna," I said, looking around the room with her, "I hope you don't mind if I finish checking things out while we talk."

"I don't mind at all," she said, opening her notebook and finding her pen. "When did you come up with idea?" she asked, anxious to begin.

"I've been designing this for about the last twenty years," I told her as we walked into the room set-up with the computers and other electronic boards I needed to use for the STEC to work. "Two of my best friends died many years ago. I knew there had to be a way to talk with them again while I was still living. So, I came up with the idea for the STEC."

"What does STEC stand for?" she asked, writing as fast as I could answer.

"STEC is short for Spiritual Tele-communication Center. I like to tell people it is like a telephone to talk to the other side."

"Oh," she said, looking up at me, "You mean it's like a cell-phone to heaven."

"That's pretty close," I said, chuckling.

"Is that where Mr. Edison will be seen at?" she asked, glancing over towards the cabinet.

"I sure hope so," I told her, walking with her to the cabinet. I turned around and saw her father staying back towards the door, letting his daughter enjoy the time alone with me for her interview. "I was just going to do a final check on things when you came in. Would you like to help me make sure everything will be alright?"

"Really?" she asked, showing her excitement of helping me. "I don't want to break your machine."

"You won't," I told her, looking back at her father smiling hard. "I need to check and make sure it is going to be dark enough in here so we can all see Thomas when he shows up. I just need you to go back and stand with your dad. When I tell you, turn out the lights so we can see how dark it will be. We need to block out all of the light, okay?"

Julianna hurried over to her dad and stood by the light switch on the wall. I walked back to the computers and told her to turn off the lights. She reached up and flipped the two switches down, turning out the lights of the room.

"Can you see any lights?" I asked, waiting for her to answer.

"There is a light coming in under the door," she said, "and the lights from your computers."

I reached into one of my boxes and pulled out a large towel. I rolled it up tightly and walked over to the door. Bending down, I rested the towel against the bottom, shutting out the light coming through.

"How's that?" I asked her, walking back to computers.

"How about your computer lights?" she asked, watching for any other lights coming into the room.

"I'll take care of those when we're ready to begin," I told her. "Now is the important part. I have to check the lights in the cabinet. If they need to be brighter or darker, I need you to tell me, okay?"

"Okay," she said, taking a couple of steps closer to the cabinet. I reached down and flipped the switch to turn on the ultra-indigo lights.

"Cool," I heard her whisper softly.

"How are they?"

"I don't know how bright they are supposed to be?" she said, hoping I could tell her if they were right or not.

I stuck my head out and looked at the lights. "They look good. Now, the next thing I have to check might scare you a little bit. If it does, I want you to tell me. If you get scared, go back and stand with your dad, okay?"

"Okay," she answered softly, letting me know she wasn't sure of what was going to happen next.

I reached over and turned the switch for the fog machine to begin working. I could barely hear the hum of the machine as I walked out to see what was happening. Soon, the fog began to fill the cabinet, creating the eerie sight I was accustomed to.

I could see Julianna watching the cabinet and begin to show a very large smile. She turned quickly to look at her dad and then back over to me.

"What is the fog for?" she asked, trying to write my answer in her notebook.

"The fog inside the cabinet will let us see Mr. Edison a little clearer," I explained to her. "Although we would be able to see and hear him, the fog just makes it a little better."

I looked down and watched as she tried hard to write what I was telling her.

"I'll tell you what, Julianna," I said, watching the fog fill the machine to create the mysterious atmosphere in the room, "Why don't you put that away for the rest of the interview. I always record these sessions that I do. Will you let me give you a copy of this one? That way, you can watch to the DVD without having to try to write down everything."

"Thank you," she said, closing her notebook. "I wouldn't want to miss anything. Maybe I could just play the DVD tomorrow instead of standing up in front of everybody and talking about it."

"I'm afraid your class would have to be about six hours long if you wanted to do that," I said, chuckling at her way of thinking. "What time is your science class tomorrow?"

"It's my third period. It starts about ten."

"Is that going to give you enough time to get your report ready?" I asked, wondering how much time she would have to get ready.

"I should have enough time."

"Alright then," I said, walking back to the computer to turn off the fog machine and start the fan to get the fog back out of the cabinet. "I think we're about ready. Tom, could you turn on the lights for me, please?"

"I just need one more thing from you," I told my assistant as the lights in the room came back on again. "Would you go and stand in front of the STEC so I can make sure the cameras are aiming in the right direction?"

Walking over to the cabinet, she turned around and smiled at her dad standing towards the back of the room.

"Okay, Julianna," I said, making a slight adjustment with the camera, "I just need you to give me a big smile so I can see how much detail this camera will give me."

"That's great," I told her from the other room. "Are you having a good time so far?"

"Yes."

"You aren't going to be scared when Thomas comes to talk to you, are you?" I asked, checking the volume of the microphones.

"No."

"Do you have a boyfriend?" I asked, trying to get her to give me more than just single word answers.

"Yes."

"What are you going to tell your class in your report tomorrow?" I asked, knowing she had to give this answer in more than just one word.

"I'm going to tell them how much fun I had and all about the STEC," she said, finally giving me a longer answer. "I hope my teacher will like it."

"I hope so too," I said, satisfied with the results I was getting from my final checks.

Julianna joined me by the computers, looking around at the different pieces of the STEC. We all turned around when we hear the door trying to open, without success because of the towel wedged at the bottom.

"I guess the towel keeps people out as well as the light, huh?" I joked, walking over and picking the towel up off the floor. I opened the door, to see Larry trying once again to come in.

"I was going to let you know the audience is beginning to show up," he said, watching the last of the fog begin to leave the cabinet. "Are you about ready?"

"I believe so," I said, winking at him slightly. "What do you think, Julianna? Are we ready?"

"Yep," she said, beaming with excitement.

"In that case," he chuckled as he started out the door, "I'll be back in about ten minutes with our guests."

Larry left, leaving Julianna, her father and me alone in the room once more.

"When they come back," I told them both, "I have to give another little talk. I would like you to stay around after the demonstration so I can give you your c.d., alright?"

We continued talking for a couple more minutes before Larry led the small group of witness's and a reporter for the local news in for the demonstration. I welcomed them as Julianna and her father walked over to an empty area of the room, trying to keep out of the way of the others while I began to explain the STEC to the others.

Soon, it would be time to start up the STEC to speak with the one who originally came up with the idea, Mr. Thomas A. Edison.

Chapter 20
Thomas Alva Edison

I was introduced to the different people Larry had invited to the demonstration. The room immediately started to fill with curiosity about the STEC. After meeting our guests, I walked to the front of the room, stopping in front of the cabinet.

I looked over at Julianna and her dad, smiling when she gave me two thumbs up, wishing me good luck. After a short ten minute presentation, I was ready to start the STEC, hoping Mr. Edison would make an appearance tonight.

"Usually, I use three different items to talk with a spirit," I told the group as I walked to the temporary computer room. I walked back out with a photograph and a pen Thomas actually used at his desk. "The three items are a photograph, a personal item and some DNA of the person. Unfortunately, I don't have any DNA of Mr. Edison, but these should be enough to allow him to connect with."

Silence filled the room as I opened the front glass door and put the items of the wooden shelf. I closed the door, making sure it was sealed and secured.

"Right now, I need to ask my assistant to get the towel and block the light from the door," I said, looking over at Julianna. She smiled and walked with confidence to the computer room, taking the towel and rolling it up tightly. She knelt down and blocked out all of the light coming in under the door.

"Before we begin," I told the group, "While the demonstration is going on, I must ask for complete silence in the room. Please make sure all of your cell-phones are turned off. I will ask Mr. Edison if he would answer a few questions, but we must wait for his answer. If we are ready, here we go."

I walked back to the computer room and began flipping switches to activate the STEC. I could hear a few of the people in the room begin to take a deep breath when the ultra-indigo lights came on and fog began to fill the machine. I double checked everything once more, making sure the microphones were working and the cameras were on.

I turned on the program for the voice speed, making sure it was running and the speakers were turned on so we could all hear our guest when he began to speak. Walking out to join the others, I smiled when I accidently brushed up against one of the people watching the cabinet.

"I'm sorry," I told him as he jumped slightly. He must have thought the spirit of Mr. Edison bumped into him instead of me.

I walked to the front of the group and began explaining a few more details of the STEC while we waited for Mr. Edison to arrive.

"Inside the STEC, there is a low tone being played. This tone is too low for us to hear, but it will attract the spirit we are looking for. As you can hear, the tone is not coming through the speakers connected to the cabinet. It usually takes between five and about fifteen minutes for a spirit to being to show itself. But, being that this is a place where Mr. Edison is acquainted with, it may not take that long."

I turned to watch as the fog began to settle, filling the cabinet with a smooth foggy look.

"When Mr. Edison begins to speak," I continued, "The computer will automatically adjust the speed of which he is speaking so we can have a normal conversation with him."

I watched as Julianna stepped away from her father and walked towards the center of the room to get a better look at what was going on in the cabinet. I looked at it, watching the fog beginning to swirl softly. Soon, everybody in the room could see a spirit beginning to take form. It would seem that Mr. Edison was making the well anticipated entrance.

Two minutes later, we were standing in the presence of one of the greatest minds in history, Thomas Alva Edison. I turned my attention away form our guest to look at the faces of the others who came in to witness the STEC at work. Perhaps the largest smile in the room was coming from my assistant, Julianna.

"Mr. Edison," I began, "It is a great honor to see you sir. My name is James Steines, the inventor of the machine you are using to speak with us tonight."

"The honor is mine, young man," he said, looking directly at me. "May I ask how you were able to accomplish this?"

"Sir," I said, wanting to make sure this was actually the spirit of our intended guest, "Before I begin, may I ask you a couple of questions?"

"I am glad to see you have thought about the verification of who you are speaking with," he said, showing half a smile back at me.

"Sir, after you finished a hard day at your business, there was a special place you went to relax. Do you remember that room?"

"Yes, I do. That would be room number twelve. Everybody knew when I walked into that room, I was to be left alone," he told us.

(Now you know the reason I enthused about the room number twelve I would be setting up the STEC in. Room 12 was Thomas Edison's secret room while he was working in Menlo Park, New Jersey.)

"We are all glad they listened to you," I said, getting ready to ask the next question to verify we were actually talking with the right spirit. "When you were married to your first wife, Mary Stilwell, you had two children. Their names were Marion Estelle and Thomas Alva Edison, Jr. You had nicknames for your two children. Do you remember what their nicknames were?"

"I haven't thought about that for many years," he said, smiling more. "I am not sure how you found out, but their nicknames were dash and dot. Since I was a telegrapher, I began calling my daughter, Marion, 'Dot', and my son, Thomas, 'Dash'."

"Thank you," I said, convinced this was the true spirit of Mr. Edison. "I hope you realize I had to ask these questions to verify who you were."

"I understand. What I do not understand is how you were able to create this machine."

I took a few minutes to explain how I came up with the idea and parts of the STEC.

"If it weren't for your genius way of thinking to find a way to allow people to have electricity," I continued, "there would have been no way I could have been able to accomplish building the machine. Thank you."

Thomas stopped looking at me when Julianna moved slightly to be closer to the front of the group. She stopped when she saw the spirit of Thomas looking directly at her.

"I see you have brought a young scientist with you tonight, James," he said, looking down at Julianna from inside the cabinet.

"Mr. Edison," I said, as my assistant walked closer towards me, "I would like to introduce you to Julianna. Julianna, this is Mr. Thomas Edison."

"It is a great pleasure meeting you, Julianna," he said, smiling softly at her. "Are you interested in science, young lady?"

"Yes, sir," She said softly. "May I ask you a question, sir?"

"Of course you may."

"Sir, in 1920, you gave an interview in the October issue of the American Magazine. In the article, you mentioned you were working on a machine to speak with spirits on the other side."

"Yes, Julianna, I was," he told her.

"My question is," she continued, "if you were working on this invention, did you have any drawings or design plans of the machine?"

"Most of the designs I was working on," he explained, "were mostly scribbled on pieces of scrap papers. I had a friend who I would sit with and talk about different aspects of creating the machine. We didn't even talk about a name for the machine if we were able to build it."

"Many people who have tried to find your plans have called it the 'TEC', the 'Thomas Edison Communicator', Julianna explained to him.

"I do not believe they will be able to find the notes I made," he told her. "As far I can see, there would no longer be a need to find the papers since James was able to invent his own communicator. Although my plans were able to speak with the spirits, I believe this machine would be much better. I would have never thought of creating a machine that would actually allow a person to see the spirit as they speak with them. I find this truly amazing. My congratulations to you, James."

"Thank you, sir," I said, suddenly feeling humbled in front of this spirit.

Chapter 21
Conclusion of the Edison Session

I glanced around the room to the others who came to witness this session with Mr. Edison. Although we have already been speaking with Thomas for almost ten minutes, I could still see by the expression on their faces they were amazed at the chance of seeing the spirit of this famous genius.

The talk with Tomas continued as he was kind enough to answer questions from the others in the room. We all stood quietly as Mr. Edison began to speak with us.

"It has truly been an honor to be able to come and speak with all of you today," he began. "I always thought it would be possible to speak with a spirit of a person who died, but I never would have thought it would be possible to see the spirit at the same time.

As you have all witnessed today, there is always ways to improve in the scientific ideas. Even though James did not realize I was working on this same project until many years into his research, all science inventors come up with their own ideas. My advice to all of you, especially you, Julianna, is to continue to follow your dreams and figure out ways to accomplish your own goals."

I glanced over to Julianna, watching her smiling when our guest spoke directly to her.

"Many people believe there is nothing left to invent," he continued. "They say this because they do not use their imagination or take the time to think of what has not been invented.

There will always be setbacks with any new invention. You must persevere and break down any barriers you come across. Eventually, the answer to your questions will appear. Do not give up on your dreams, or your inventions, but keep on trying. You will succeed."

I kept quiet as I realized what he was speaking the absolute truth. There were many setbacks and brick walls I had to break through with the STEC. I looked over and watched Julianna listening intensely to every word he was saying.

"I must be leaving soon," Thomas told the group. "It has been my pleasure to be called to speak to you. James, if you would like, please call for me again and I will be happy to come and speak with you at any time."

"Thank you, sir," I said, surprised at his kind offer. "Since I have your vibration numbers, I will talk with you again."

"Julianna," he concluded, "I hope you will continue your schooling and follow your own dreams. Nobody can guarantee your success except for you and you alone. It seems you have found a new friend in James. I hope he will be kind enough to help you if you need to ask him."

Julianna looked up at me, watching as I shook my head yes that I would help her if she needed it. She smiled as we both turned back to listen to Thomas end the session before disappearing from the cabinet.

"It is now time for me to leave," Thomas told us as the fog began to swirl slightly. "As I said earlier, it has been an honor to come and speak with you. I wish you all the best in your futures. With that, I shall take my leave of you all. Thank you again, James, and congratulations on your success."

"Thank you, sir," I said, watching the fog beginning to swirl more when the spirit of Mr. Edison started to leave the cabinet. Soon the cabinet would be empty, leaving only the fog as it began to slow down,

I weaved my way to the back of the room, reaching over to turn on the ceiling lights. I looked at the people in the room as they stood quietly, still absorbing the fact that they actually listened to the words of man who died over eighty years ago.

Keeping the computers and the cameras running, I walked back to the area between the cabinet and the invited group of people. I was stopped by each one of them as they stretched out their hands to congratulate me for the success of the STEC.

"I would like to thank you all for joining me tonight while we talked with Mr. Edison," I told them all, looking at the many faces listening to me. "If you would, I would ask each of you to give Larry your address so I can send you a copy of the session you were just a witness to. Do any of you have anymore questions you would like to ask before we call it a night?"

"Who are you going to talk to next?" a gentleman in the middle of the crowd asked.

"For right now," I told him, "I think it would be difficult to top this session. I have already spoken with the three spirits I built the machine to contact. Let's see what the future holds."

"When are you going to begin producing the STEC for sale to the public?" another person asked.

"I am in the process of talking with a few companies who have shown interest in the STEC, but I haven't made any decisions yet. Would any of you purchase one if they became available?"

Every person in the room said yes that they would buy one for their own private use. We spoke for a few more minutes before Larry led the group out of the room after thanking me for allowing them to be a part of the historical demonstration.

After everyone left, I turned around to find Julianna and her dad standing in front of the cabinet talking to each other.

"What do you think?" I asked her.

"That was great," she said with excitement in her voice. "I can't thank you enough for inviting me and Dad to be here tonight. I will never forget this night."

"You are very welcome," I told her, not understanding how a young woman could smile as big as she was. "I hope the demonstration tonight will help you in your report tomorrow for your science class."

"It will," she said. "I just don't know what parts of tonight I will have to leave out of the report. If I don't leave something out, I might have to take three days to finish the report."

"I'm sure you will figure out something," her dad told her. "You always do."

"Dad says I think too much," she joked. "I told him I don't think enough."

"It sounds to me like you think just enough to get the problem solved," I told her, watching her continue to smile. "I hope you never stop thinking like you do. Something is telling me that someday I will be coming to one of your demonstrations of your invention."

"You will be the first one I will invite," she joked.

"Before I forget, I have to follow Mr. Edison's instructions," I told her, reaching into my shirt pocket and pulling out one of my business cards for her. "Here is my card and telephone number. Thomas said I was to help you if you ever had a science question you needed help with."

Her dad took the card and put it in his pocket, making sure Julianna wouldn't accidently misplace it. She gathered her notebook and purse from the computer room, before her and her dad left the room after a great experience with the STEC.

Larry came back to the room to escort my final two guests back to the front door. Julianna shook my hand firmly, thanking me for a great demonstration. Her father did the same before walking out the door with Larry.

A few minutes later, I began my final preparations to save the information from the session with Thomas. Larry came back and talked with me for a few minutes, thanking me for the demonstration and congratulating me once more for my success with the STEC.

Chapter 22
Julianna's Science Class

After talking with Larry, I was allowed an extra two days to pack up the STEC for transport back home. Arriving early the next day, I began taking down my invention, putting the miscellaneous pieces in their proper moving boxes.

I was almost ready to take down the computer system when I remembered I had yet to make a copy of last night's session for Julianna. Finding a blank DVD disk, I quickly made a copy for Julianna to share with her science class. I would wait until I got back home to make the copies for the others who witnessed the demonstration last night.

Looking at the time, I knew I could call Joseph to have him take me to Julianna's school so I could deliver it to her in person at her science class. An hour later, Joseph pulled his limousine in front of the business doors to take me to school.

Last night while we were getting the STEC ready for the demonstration, Julianna told me her science class began about ten. I talked to Joseph when he picked me up and he assured me we would have plenty of time to get there before her class began.

We made it safely to the school, drawing the attention of a few students looking out of the windows at a long black limo pulling up to the front of their school. Before I could get out, most of the windows had students looking out, trying to get a glimpse of who was going to get out of the car.

Joseph opened the door for me, letting me know he would be back in an hour to pick me up. Walking into the front door of the school, I looked for the sign that would lead me to the main office. Finding the office, I walked in and asked to speak with the high school principal, Mrs. Gleason.

After meeting with Mrs. Gleason, she offered to take me to the classroom where Julianna was getting ready to give her report on being with me last night during her visit. I stayed outside as Mrs. Gleason went in and asked for her teacher to come out in the hall for a minute.

Julianna's science teacher, Mr. Snelling, came out to the hallway to meet me. I tried to keep out of the sight of Julianna, but was able to see her standing in front of the class with her note-cards in hand.

Mrs. Gleason introduced me to Mr. Snelling, letting him know I was there to bring the DVD from last nights demonstration of the STEC for one of his students. He asked if I could come in for a few minutes and talk with his class about the STEC.

"I'd be more than happy to," I joked with him, "as long as you give Julianna an 'A' on this report of hers."

He chuckled and smiled as he agreed to my terms, telling me she was just beginning her report. He walked inside the classroom, stopping Julianna from continuing with her report and motioning for her to step outside to the hallway with him.

Looking confused at why she was asked to stop, she walked to the back of the room, following her teacher outside. I could see from the big smile that was forming, she was happy to see me at her school.

"Hi, Jim," she said, reaching out her hand to me. "What are you doing here?"

"Hi, Julianna," I said, shaking her hand. "I forgot to give you your DVD of the session from last night. I hope you don't mind me coming here today to give it to you, do you?"

"Thank you," she said, taking the DVD, looking at it carefully. "Can you stay and listen to my report about the STEC?"

"I would love to," I said, wondering if her large smile was ever going to disappear. "If you would like me to, I would be happy to answer any of the questions the others might ask if you don't have the answers, alright?"

I turned around and thanked Mrs. Gleason for showing me to the classroom as Mr. Snelling and Julianna opened the door for all of us to go back inside. I stood by the door, letting it close behind us as Julianna went back and stood in front of the class.

The other students took their eyes off of me as Julianna began her report again.

"Before I start," she said, pointing over to me, "I would like to introduce Mr. James Steines. He is the inventor of the STEC machine. STEC stands for the Spirit Tele-communication Center. This machine lets a person talk with the spirit of somebody who died."

The students looked over at me, a few of them jotting down notes on the papers in front of them.

"Yesterday," Julianna continued, "I went with my dad to listen to Mr. Steines talk about his invention. I asked him if I could interview him for this report today. Instead of just giving me an interview, he asked me if I wanted to see the demonstration. This DVD is a copy of the session I was invited to go to."

She held it high for the entire classroom to see. Setting the DVD down on her teacher's desk, she continued with her planned report.

"Many years ago Mr. Steines lost two of his best friends. Wanting to talk to them again, he thought of the idea of building a machine that could talk to the spirits of people who died. He called this machine STEC. STEC stands for the 'Spiritual Tele-communications Center'. It took him over twenty years to figure out how to create the STEC.

Finally, about a year ago, he was able to build a working machine. I know it works because I saw it working last night. A couple of months ago, Mr. Steines was invited to come here to give a demonstration to a few people.

I was excited when I was actually able to see the spirit of Thomas Edison appear in the cabinet of the STEC and talk with us. I was even given the chance to ask a question to Mr. Edison. Since he was planning on building the same type of a machine to talk with spirits back in the 1920's, I asked him why there weren't any drawings of his machine anywhere.

He looked right at me and said that there was, but they haven't been found yet."

Julianna stopped suddenly and looked at me in the corner of the room watching her. She focused her attention to Mr. Snelling in the back of the room and asked him a question.

"Mr. Snelling," she began. "I know this is my assignment, but the end of the class is coming soon. Could we listen to Mr. Steines for the rest of the class? If you would let me, I can finish this tomorrow. Maybe we can watch the DVD so everybody can see the demonstration from last night."

"I think you should ask Mr. Steines if he would be willing to speak with the class first, don't you?" he said, knowing I already agreed to it.

"Is that okay with you, Jim?" she asked, smiling over at me.

"I would love to."

"I would like you all to meet Mr. James Steines," Julianna said, looking back towards her classmates.

I was welcomed by a small round of applause from Julianna's tenth grade science class as I walked over to join her.

"May I ask you to stay up here with me?" I asked my new friend. "I might need some help answering their questions."

"Sure," she said, smiling back at me.

"Hi, everybody," I began, watching the eyes of the students beginning to look up to me. "I know we don't have much time left before your next class, so I'll talk for just a few minutes and will answer some of your questions."

Chapter 23
Many Good Questions

After giving a short ten-minute talk about the STEC, I offered the students the opportunity to ask questions if they wanted. I was surprised at the level of questions coming from this tenth-grade science class.

Most of the students began raising their hands, hoping I would answer their questions. I pointed to a young man towards the center of the class, waiting for his question.

"You mentioned that each spirit has its own different vibration. What type of calculations did you use to figure out the different vibrations of the spirits?"

"As time continues," I began to explain, "the vibrations of time itself changes. When were you born?"

"May 21st, 1999," the young man answered.

"May I use the blackboard, Mr. Snelling?" I asked, hoping I could give the class a little demonstration of how I came up with the right vibrations.

"Of course," he answered, interested in what I was showing his students.

"Let's say you were born at eleven thirty-seven and seventeen seconds on the 21st of May in the year 1999," I said, writing down the date on the blackboard. Even though I didn't have my computer to use the calculations, I could do them in my head. I began doing the math on the blackboard so the class could see. "Using a formula I devised, I can figure out exactly what the vibration of the past was when you were born."

Looking back, and double checking my calculations, I wrote the final answer on the board for him.

"The vibration I would begin looking for your spirit would be this number, right here. Since I know the vibration number when you were born, there would be no reason for me to look any earlier to bring in your spirit for a talk."

I watched the young man writing down his vibration point, double checking the numbers on the blackboard with the ones he wrote down.

"The spirits I try to contact are those of the people who have died," I continued, walking back to the front of the desk. "I don't feel comfortable trying to contact the spirit of a person who is still living. I'm not sure what would happen to that person if I were to try."

I pointed to a young lady on the left side of the room, who had raised her had earlier.

"Do you have a question?" I asked her.

"Weren't you afraid the first time you saw a spirit?" she asked softly.

"I think I was more excited than scared," I told her, chuckling under my breath. "I had seen spirits before when I was visiting a cemetery back home to do some research, so I was comfortable with them. Some people might be afraid, but that's because of what they have been told about spirits."

I looked over at Julianna, enjoying her time listening to me.

"Last night," I continued, "I had a very special assistant who was helping me out with the demonstration. This was also her very first time seeing a spirit and being able to talk with him. I think she would be the one to ask about being scared."

"Were you afraid last night, Julianna?" the same young lady asked.

"I was more nervous than afraid," she said, looking over the class. "I don't know if I was more nervous knowing I was going to see Thomas Edison, or if it was because I was asked to help Jim."

I looked over at her and just smiled.

"When the lights went out and the cabinet lit up," she continued, "It was a very strange feeling. Then, Mr. Edison began to show himself to us. It took a few minutes, but when he did, I was amazed at what I was seeing. His spirit looked just like the pictures we have all seen of him.

I can never thank Jim enough for asking me to assist him last night."

"It looks like we might have enough time for one more question," I told everyone as I glanced up at the clock on the wall in the back of the room. Many others raised their hands, hoping I could answer their question before the bell rang.

"I have a question for you, Mr. Steines," the teacher asked from the back of the room. "What kind of an education do you have? Did you go to college to learn what you needed to know to build your invention?"

"Unfortunately, no," I said, knowing where he was going with his question. "I didn't go to college after I graduated high school. Looking back, I wish I would have. When I decided to build the STEC, I did my research on the computer and taught myself the math and science I would need.

If I would have gone to college, I'm sure it wouldn't have taken me as long to build the STEC as it did."

"What are you going to build next?" Julianna asked, looking over at me.

"I'm not sure yet," I joked with her. "Do you have any ideas?"

The loud bell rang, telling me it was the end of the class period. The class applauded as Mr. Snelling thanked me for talking to them today. Julianna was the last one to leave, hoping she could spend a couple more minutes with me before her next class began.

"Before I leave," she asked, picking up her DVD from her teachers desk, "Could you please sign this for me?"

I took out my pen and pulled the front cover out of the plastic case to sign for her. She smiled when she read the words I wrote for her.

'Julianna, thank you for your help. J. W. Steines'

We left the classroom when other students began filing in for their time with Mr. Snelling. I walked over and thanked him personally for allowing me the time to speak with the class. He thanked me also, assuring me that Julianna would be getting an 'A' for her report.

Julianna and I walked out of the class, quickly saying good-bye before she started walking to her next class before the bell rang again. Walking towards the front door, I saw the principal Mrs. Gleason standing in the hall in front of the main office.

"Thank-you for letting me talk to Mr. Snelling's class today," I said, shaking her hand.

"It was a pleasure meeting you," she said, shaking my hand. "Something tells me you made a young lady very happy today. I heard they are going to play her recording tomorrow in class. I might have to visit the class to watch it with them."

"I hope you enjoy it," I said, watching her smile and keeping an eye on the rest of the students hurrying to their classes. We said good-bye and I walked towards the front doors of the school, hoping Joseph would be there with the limo to take me back to finish packing up the STEC for transport back home.

I saw Joseph waiting for me, just like he said he would be. He reached down and opened the back door for me, as many eyes were peeking out of the windows once more, trying to see who the lucky person riding in a limousine was. I turned around and waved at them, smiling as they started waving back at me.

Joseph closed the door and walked around to the driver door, opened it and sat down behind the wheel.

"Was she surprised?" he asked as we drove away from the school.

"Yes, she was. She did a great job at the beginning of her report, then I was asked to speak with the class. We had a great time."

We continued talking all the way to the building where my STEC was waiting to be packed up again. I thanked him as he stopped in front of the steps leading inside the building. He asked me to call him when I was ready to leave for the day, or if there was anything else I needed.

Walking into room twelve reminded me of the conversation with Mr. Edison when he explained to us all what that room meant to him. I opened the door, looked around and saw that everything was just as I left it a couple of hours ago.

Looking at the work I had ahead of me made me think it would take a while, even if I had to skip lunch. Walking into the temporary computer room, I chuckled when I saw a note with a three musketeers candy bar sitting on the desk. I picked up the note and chuckled more as I read the words.

'*Just in case your stomach starts growling. Larry.*'

I began packing the pieces of the STEC, making sure each piece was in the proper box. Soon, I had everything ready for the long trip back home. Larry came into the room, just as I was loading up the boxes and helped me look around the room to make sure I wasn't missing anything.

Larry invited me out for dinner before I went back to the hotel for the night. I accepted and enjoyed a great meal at one of the local restaurants with my host. I called Joseph to tell him I wouldn't be needing a ride tonight and I would see him in the morning so he could bring me here to finish packing the STEC for transport back home.

Tomorrow, it would be time to leave this busy town after a very successful demonstration and go back to the slower and more relaxing lifestyle I have become accustomed to.

Final Chapter
Who Would You Talk To?

I hope you enjoyed the book. As I mentioned earlier, the proceeds of the sales of this book will allow me to continue working on the STEC. With your help of the purchase of this book, the possibility of finishing the STEC and making it available to the public could happen sooner than expected.

Even though I haven't been able to talk directly with a spirit yet, I am confident that I will be able to soon. With all of the spirits I want to speak with, it would be difficult to think of the first spirit to call.

Of course, I would love to have the chance to talk with my two best friends who I lost many years ago, as well with my grandpa again, as I still miss them all very much. Then again, I would find it interesting to speak with people of history.

I wouldn't mind having a discussion with Abraham Lincoln, or maybe even a member of history like an ancient king or queen from the past. My concern about talking with the people of history is what would happen if I found out the history we know, and were taught, is incorrect?

The history of this earth was written by men, sometimes hundreds of years after the event they wrote about. Imagine how great it would be to actually speak with the spirit of a person who the history was written about.

Imagine being able to use the STEC to meet and talk with a person who lived through the middle ages in England. Or, be able to talk with one of your ancestors who died a few hundred years before you were born.

The possibilities of who you could talk with could be endless, if you have the right tools to contact them with. The further back in time you want to find a spirit to talk with, the less chance you will be able to find a DNA sample or an object owned by that person. Obviously you wouldn't be able to find a photograph of them to use, unless they lived after 1837 when the photograph camera was invented.

The best chance of being able to speak with a spirit before 1837, would to have an object or some DNA from that person. I understand how difficult it would be to acquire some of the persons DNA, yet it would be possible.

For instance, an archaeologist can retrieve the DNA from a skeleton they may find during an excavation site. Using the translator program, it would be possible to carry on a conversation with a spirit from anywhere in the world.

I felt it was important to include the story of Julianna to promote the necessity of education of young people. These young students are the future of this world we are living in.

Technology is advancing at a faster rate than ever before and the world will be needing brilliant minds to keep up with the demand. Students must be excited about what is going to happen in the future. The imagination of these students must not be held back.

If a student follows the three words I was told many years ago, 'Conceive it, Believe it and Achieve it', they must be allowed to follow through with their dreams. If allowed to follow their dreams, the youth of today can become successful in anything they do and create a brighter future for humanity.

Combining the knowledge of the great minds of the past using the STEC with the minds of today will allow the inventors of today to design and create new machines that nobody has ever thought of before. With these new inventions, the future of the earth could suddenly look a lot brighter.

www.ingramcontent.com/pod-product-compliance
Lightning Source LLC
Chambersburg PA
CBHW051315170526
45166CB00002B/548